MASTERS OF
THE PLANET

MASTERS OF THE PLANET

THE SEARCH FOR OUR HUMAN ORIGINS

IAN TATTERSALL

St. Martin's Griffin   New York

www.stmartins.com

Designed by Letra Libre, Inc.

The Library of Congress has cataloged the hardcover edition as follows:

Tattersall, Ian.
    Masters of the planet : the search for our human origins / Ian Tattersall.
        p.   cm.
    Includes bibliographical references and index.
    ISBN 978-0-230-10875-2 (hardcover)
    1. Human beings—Origin. 2. Human evolution. 3. Evolutionary psychology.
I. Title.
    GN281.T364   2012
    599.93'8—dc23

                                                                2011034415

ISBN 978-1-137-27830-2 (trade paperback)

Our books may be purchased in bulk for promotional, educational, or business use. Please contact your local bookseller or the Macmillan Corporate and Premium Sales Department at 1-800-221-7945, extension 5442, or by e-mail at MacmillanSpecialMarkets@macmillan.com.

First published by Palgrave Macmillan, a division of St. Martin's Press LLC

First St. Martin's Griffin Edition: May 2013

D   11

For Gisela, Tat and Chub

CONTENTS

CONTENTS

MAJOR EVENTS IN HUMAN EVOLUTION

Event	Thousand Years Ago
Origin of Life	3,500,000
Origin of Primates	60,000
Group containing humans and apes begins to diversify	23,000
Earliest hominids (bipeds) appear in Africa	7–6,000
First *Australopithecus*	4,200
Earliest possible use of sharp stone for cutting	3,400
Beginning of glacial cycle	2,600
Distinct expansion of grassland fauna in Africa	2,600
Earliest documented manufacture of stone tools	2,600–2,500
Claimed "early *Homo*" fossils appear	2,500–2,000
First *Homo* of modern body proportions in Africa	1,900–1,600
Hominids first leave Africa (Dmanisi)	1,800
First stone tools made to deliberate shape	1,760
Homo erectus appears in Asia	1,700–1,600
First *Homo* fossils in Europe	1,400–1,200
Earliest evidence of domesticated fire in hearths	790
Homo antecessor appears in Europe	780

First Old World–wide hominid, *Homo heidelbergensis*	600
First evidence of Neanderthal lineage in Europe	> 530
Earliest blade tools in Africa	500
Earliest wooden spears, hafted tools	400
First evidence of constructed shelters	400–350
Earliest prepared-core tools	300–200
Origin of anatomically recognizable *Homo sapiens* in Africa	~200
First possible beadwork	~100
Earliest engravings, heat-treatment of silcrete	~75
Exodus of cognitively symbolic *Homo sapiens* from Africa	70–60
First modern humans in Australia	60
First modern humans in Europe, flowering of art and symbols	40–30
Extinction of Neanderthals, *Homo erectus*	~30
Homo floresiensis extinct	14
Last Ice Age ends	12
Plant cultivation and animal domestication begin	11

PROLOGUE

Stare at the face of a chimpanzee. Look deep into its eyes. Your reactions will almost certainly be powerful, complex, and murky. Perhaps on balance you'll want to recoil, as the Victorians tended to, perceiving in the apes a bestial savagery that served as an unwelcome reminder of humanity's feared and (usually) repressed dark side. In our own day, though, you'll much more likely see in the chimpanzee something more positive: not a failure to achieve human status, but an inchoate glimpse of the deep biological foundations on which our modern civilization and creativity are ultimately based. Still, whatever your exact reaction may be, it will certainly come from perceiving a lot of yourself in those eyes—and the side of the human coin you will see reflected will depend entirely on you, not on the chimpanzee.

This ambiguity makes it very frustrating that the chimpanzee can't articulate his state of mind to us, or answer our questions about it. But then, for all of his physical differences, if he *could* talk he would be one of us. Nothing else he could do would place him more emphatically in the human camp, for it has been recognized since ancient times that language defines us as nothing else does. Indeed, the Scottish jurist James Burnett, Lord Monboddo, anticipated evolutionary thought as early as the 1770s when he suggested that the acquisition of language was the key feature that had levered humankind away from the "lower" animals: an intuitively attractive notion that has been revisited by numerous thinkers since. During the quarter millennium that has elapsed since Monboddo wrote, a vast trove of information bearing on this issue has accumulated, in numerous areas of science that range from linguistics through genomics to neurobiology. Most importantly, we have learned a great deal about the diversity and behaviors of our precursors on this Earth: certainly enough to allow us to begin speculating with some confidence about how, when, and in what context humankind acquired its extraordinary habits of mind and communication.

The story of how we became human is a long one, and it is one that is best recounted from its ancient beginnings, well before there was any firm hint of what was to come. So let's return for a moment to that chimpanzee and its relatives. It's hardly surprising that the apes are so unsettlingly like us. They are our closest living relatives in the biosphere, sharing with us an ancestor that lived perhaps as recently as seven million years ago—a mere eye-blink in the history of Life. But in that short time no other animal lineage has changed nearly as much as ours has. This means that even though they, too, have changed, we can reasonably look to chimpanzees and their relatives for clues as to what our common ancestor was like. And if these primates serve as a reliable guide, that ancestor was an extremely complex creature indeed. Chimpanzees bond, quarrel, and reconcile; they deceive; they murder; they make tools; they self-medicate. They live in hugely complicated societies; and in the struggle for status within those societies they form intricate alliances, and indulge in intrigues that some observers have described as nothing less than "politics." If humans had never evolved, apes would almost certainly be today the most cognitively complex animals that had ever existed.

Yet here we are. And the story of how we got here from there, leaving our ape relatives in the dust (or at least in the trees), is perhaps the most intrinsically fascinating and complex story that our narrative-loving species has ever tried to tell. But at the same time it is an elusive one. For while comparing ourselves with apes may help us establish a starting point for our long evolutionary trajectory, it turns out that we modern human beings are not simply an improved version of them. Instead, we are an altogether unprecedented presence on our planet; and explaining the unique has always been a thankless task.

Despite the difficulties inherent in trying to explain ourselves, we have a solid foundation on which to start. The past century and a half has witnessed the accumulation of a remarkable fossil record that, although it will never be complete, already gives us a substantial glimpse of the appearances and astonishing diversity of those ancestral and collateral relatives who preceded us. What's more, these human precursors are unusual in having left behind an archaeological record—butchered bones, stone artifacts, living sites—that speaks eloquently of their daily activities, and of how those activities became more complex as time progressed.

Documenting the huge physical and technological changes that accompanied the long trek from ancient ape to modern human is, at least in principle, a relatively straightforward task. But the secret to the particular kind of success our species enjoys today lies in the very unusual way in which our brains handle information. And mindset is something that is very hard to read from bones or material leavings, at least up to the point at which we have overwhelming evidence for the presence of an intellect equivalent to our own. What is evident, though, is that this final point was reached very late in time—at least compared to the earliest appearance of the human family, although in modern historical terms it was dizzyingly early. Many may find this tardiness rather surprising, because traditionally we have been taught to view the long human story as an extended and gradual struggle from primitiveness toward perfection—in which case, we might anticipate finding early harbingers of our later selves. The reality, however, is otherwise, for it is becoming increasingly clear that the acquisition of the uniquely modern sensibility was instead an abrupt and recent event. Indeed, it was an event that took place *within* the tenure on Earth of humans who looked exactly like us. And the expression of this new sensibility was almost certainly crucially abetted by the invention of what is perhaps the single most remarkable thing about our modern selves: language.

This final communicative and cognitive leap is far from the whole story. The underpinnings of the modern body and mind reach far back into the past, and most of this book is devoted to examining the deep foundations on which the amazing human phenomenon was built. For nothing of what we are today would have been possible in the absence of any aspect of our unique history. And although it is in Africa that we find the earliest stirrings of the modern mind, the vagaries of the record are such that it is only when we contemplate the astonishing cave art of Ice Age Europe that we encounter the first evidence of human beings who not only thought as we do, but who left behind an overwhelmingly powerful body of evidence to prove it.

SYMBOLISM AND THE ART OF THE CAVES

Best exemplified by the famous animal images on the ceilings and walls of caves such as Spain's Altamira and France's Lascaux and Chauvet, the raw power and sophistication of this ancient art is somehow magnified by the knowledge that its painters lived in an unthinkably remote epoch of modern human history. For, despite their brilliance in color and concept, these extraordinary works were the product of hunter-gatherers who lived around the peak of the last Ice Age, between about thirty-five thousand and ten thousand years ago. These were harsh times of cool summers and bitterly long winters, during which trees were often almost entirely banished from a landscape that is thickly wooded today. The antiquity of this art is astonishing; but exposure to it nonetheless makes you fully understand Picasso's alleged remark that the Ice Age painters had left him little to accomplish. Certainly, it's impossible to imagine better evidence that the wonderful and unprecedented human creative spirit was already fully formed at that distant point in modern prehistory.

This realization had not come easily. Intuitively, it was difficult for nineteenth-century scientists to accept that the ancient Ice Age inhabitants of southern France and northern Spain had created an artistic tradition—embracing painting, engraving, sculpture, and bas-reliefs—that, at its best, had equaled or even exceeded in its power anything achieved since. After the first (and among the finest) cave paintings were discovered at Altamira in 1879, immediate admiration rapidly gave way to doubts. How could such refined and accomplished art possibly be the work of hugely ancient people? How could it have been produced by "savages" without fixed abode: mere hunters and gatherers who had roamed the landscape and availed themselves of its bounty, quite the antithesis of civilized nineteenth-century folk who worshiped in magnificent cathedrals, built sturdy houses for shelter, and put the land and what grew on it to work for them? It took repeated discoveries of ancient art, in virgin caves and at untouched archaeological sites, to convince the world that

*Monochrome rendering of a now badly faded polychrome wall painting,
probably around 14,000 years old, in the cave of Font de Gaume, France. A
female reindeer kneels before a male that is leaning forward and delicately
licking her forehead. Drawing by Diana Salles after a rendering by H. Breuil.*

you could indeed have both a sophisticated mind and a "primitive" lifestyle: to make acceptable the notion that, those many tens of millennia
ago, people had existed who did not live in houses and till the fields, but
who nonetheless made fabulous art, led mysteriously complex lives, and
were just like us in all their cognitive essentials.

Of course those ancient people, and the larger societies whose beliefs
and values those images at Lascaux and Altamira embodied, vanished
long ago. So, although we have at our disposal miraculously preserved
material evidence of the creative spirit of those long-gone humans, we
will never know for sure just what those beliefs and values were. Nonetheless, for all their cultural and temporal remoteness, we *can* be secure
in the knowledge that those ancient people of Altamira and Lascaux and
elsewhere were *us* in all essentials, imbued with the same remarkable
human spirit that animates us today.

Significantly, the walls of Lascaux and other caves are not decorated only with animal images, drawn with the deftness, observation,
and clever stylization that place their creators among the greatest artists
ever. Among and upon those instantly recognizable animal figures, the
artists placed geometric motifs—grids, lines of dots, dartlike signs—that

clearly had very specific meaning to their creators. Sadly, today we have no way of knowing just what it was the artists had intended to express; but if you consider the clear specificity of the images together with the complex ways in which they are juxtaposed, you rapidly begin to realize that this art was not simply representational. It was *symbolic*. Every image in this cave and others, realistic or geometric, is drenched with meaning that goes far beyond its mere form.

Even though we can't know exactly what the art of Lascaux meant either to its creators or to those for whom it was intended (whether the two were the same, we'll never be certain), what is undeniable is that this art signified something well beyond what we are able to observe directly. And this is, oddly enough, one of the most powerful of the many reasons why so many of us resonate to Ice Age art at the most profound of levels. Because, for all the infinite cultural variety that has marked the long road of human experience, if there is one single thing that above all else unites all human beings today, it is our symbolic capacity: our common ability to organize the world around us into a vocabulary of mental representations that we can recombine in our minds, in an endless variety of new ways. This unique mental facility allows us to create in our heads the alternative worlds that are the very basis of the cultural variety that is so much a hallmark of our species. Other creatures live in the world more or less as Nature presents it to them; and they react to it more or less directly, albeit sometimes with remarkable sophistication. In contrast, we human beings live to a significant degree in the worlds that our brains remake—though brute reality too often intrudes.

Human beings are unusual in many ways, physical as well as cognitive. But our unique mode of processing information is without any question the element that, more than any other, marks us off as *different* from other creatures; and it's certainly what makes us *feel* different. What is more, as I hope this book will convince you, it is entirely without precedent. Not only is the ability for symbolic reasoning lacking among our closest *living* relatives, the great apes; such reasoning was apparently also absent from our closest extinct relatives—and even from the earliest humans who looked exactly like us. At the same time, we modern humans have a huge amount in common intellectually with all of those relatives, vanished and living; and, even more to the point, no matter how much we may vaunt our rationality, we are most certainly not entirely rational beings: a point that should need no belaboring to any observer of our species. One major reason for this is that, through the vagaries of a long and eventful evolutionary history, some of the newest components of our brains—those strange, complex organs in our heads that govern our behavior and experience—communicate with each other via some very ancient structures indeed.

Because of the peculiar construction resulting from their complex history, our brains are far from directly comparable to a feat of human engineering. Indeed, they are probably not comparable at all. For engineers always strive, even where they are consciously or unconsciously constrained, for *optimal* solutions to the problems they are facing. In contrast, during the long and untidy process that gave rise to the modern human brain, what was already there was always vastly more influential on the historical outcome—what actually *did* happen—than any potential for future efficiencies could be. And thank goodness for that. After all, if our brains had been designed like machines, if they had been optimized for any particular task, they would *be* machines, with all of the predictability and tedious soullessness that this would imply. For all their flaws, it is the very messiness and adventitiousness of our brains that makes them—and us—the intellectually fertile, creative, emotional, and interesting entities that they and we are.

This perspective conflicts with the view of evolution that most of us were taught in school—where, if it was mentioned at all, this most fundamental of biological processes was usually presented as a matter of slow, inexorable refinement, constantly tending toward achieving the perfect. So, before we embark on the human story, it seems reasonable to take a few moments to look more closely at the remarkable process that operated to produce us—because, extraordinary as we may justifiably think ourselves, we are actually the result of a perfectly ordinary biological history.

THE VAGARIES OF EVOLUTION

Let's start right at the beginning, with the overarching pattern in which Nature is organized, because this is the clearest tip-off we have to the mechanisms lying behind our appearance on the planet. There is a clear order in the living world. The way in which the diversity of animals and plants around us is structured is not haphazard in the least. Instead, it shows an across-the-board pattern of groups within groups. Among the mammals, for example, human beings are most similar to apes; the apes and humans together are most similar to the monkeys of the Old and New Worlds; and apes, humans, and monkeys all resemble lemurs more closely in their anatomies than they do anything else. Jointly, these primates form a distinctive cluster within Mammalia, the order that groups together all the warm-blooded, furry animals that suckle their young. All mammals in turn belong to a bigger group known as Vertebrata (the backboned animals—fish, amphibians, reptiles, and birds, as well as mammals), and so on.

Every other organism is similarly nested into the living world; and graphically, this pattern of resemblance is best expressed in the form of a repeatedly branching tree. Ultimately, every one of all the many millions of living organisms can be embraced within one single gigantic Tree of Life. In this greatest of all trees, biologists group the tiniest branch tips (species, e.g., *Homo sapiens*) into genera (e.g., the genus *Homo*), which are in turn grouped into families (Hominidae), which are grouped into orders (Primates), and so on. As you move up the tree, each successive level departs farther in its configuration from the common ancestral form at the bottom, and from equivalent neighboring branches. And although it is possible to study this self-evident Tree of Life in purely structural terms, the most interesting thing is to know what caused it.

The only testable (and thoroughly tested) scientific explanation of this pattern of resemblance is common ancestry. The similarities that clue us in to the shape of the tree are inherited from a series of shared ancestral forms, whose descendants have diverged from them in various respects. Similar forms share a recent common ancestor, while more disparate ones shared an ancestor much more remotely in time—allowing differences to accumulate over a longer period. No matter how dissimilar they may now appear to the eye, all life forms are ultimately linked at the genomic level to a single common ancestor that lived more than 3.5 billion years ago.

The nineteenth-century naturalists Charles Darwin and Alfred Russel Wallace were the first to come up with a convincing mechanism by which divergence from a common ancestor could occur. Darwin called this instrument of change "natural selection." Once pointed out, this natural process seemed so self-evident that Darwin's famous contemporary Thomas Henry Huxley publicly berated himself for his own failure to think of it. In a nutshell, natural selection is simply the preferential survival and reproduction of individuals that are better "adapted" to their environments than their fellows, in features inherited from their parents. And it is pretty much a mathematical consequence of the fact that, in all species, each generation produces more offspring than survive to reproduce. The idea here is that, over enough time, those with more advantageous inherited characteristics will have greater reproductive success, and therefore will nudge the population in the direction of better adaptation. In this way, members of the lineage will change in average appearance and ultimately evolve into a new species.

That was the theory, anyway, though it has subsequently been noticed that natural selection may well act mostly to trim off both extremes of the available variation, keeping the population more or less stable. Another complication is that, when we think of adaptation, we usually have in mind one single anatomical feature, or behavioral property, of the animal in question: the structure of its foot or pelvis, say, or its "intelligence." Thinking of just one feature in isolation, it is easy to envision how that structure might have been improved over time by natural selection. Yet we now know that all organisms are astonishingly complex genetic entities, in which a remarkably small number of structural genes (exactly how many we humans have isn't known for sure, though most current bets are in the 23,000 range) govern the development of an enormous number of bodily tissues and processes. In the end, natural selection can only vote up or down on the entire individual, which is a real mash-up of genes and of the characteristics they promote. It cannot single out specific features to favor or disfavor.

This, though, blurs the "fitness" picture. It is, for example, of little value to be the smartest member of your species if, in an environment crawling with predators, you are also the slowest—or even just the most unfortunate. What's more, in an indifferent world your reproductive success may not in the end have much to do with how magnificently you are adapted to any one thing. Whether or not that predator gets you, or whether or not you get the girl, may simply be a function of blind luck and circumstance. The upshot of complications such as these is that evolutionary histories, certainly as we see them reflected in the fossil record, are not produced by the reproductive fates of individuals alone. Indeed, in a world of constantly changing environments, and of ceaseless competition among different kinds of organisms for ecological space, it is more often the fates of entire populations and species that determine the larger evolutionary patterns we observe when we look back at the fossil record.

And there are yet other reasons for not expecting that evolution should produce tidy perfection. As I've already suggested, change can only build on what is there already, because there is no way that evolution can conjure up *de novo* solutions to whatever environmental or social problems may present themselves. As a result, we are all built on modified versions of a template ultimately furnished us by an ancient ancestor. History severely limits what you can potentially become not simply because you must necessarily be a version of what went before, but because genomes, dedicated as they are to the propagation of mind-bogglingly complex systems, turn out to be hugely resistant to change. In fact, they provide the ultimate example of "if it ain't broke, don't fix it." After all, fiddling around with anything as intricate as a genome is asking for trouble: most random changes to a functioning system this complicated simply won't succeed. The fact that changes in the genetic code carry huge risks explains the inherent conservatism of genomes. It also explains why some organisms that look hugely different to the eye have amazingly similar genes: I've heard it said that we share over 40 percent of our genes with a banana, while a gene that is highly active in determining human skin color is also responsible for regulating the dark stripes on the side of a zebrafish.

It may seem amazing that the same genes or gene families can influence structure across a spectrum of organisms that look as vastly different as, say, a human being and a fruit fly. But it makes sense when you consider not only that all organisms share an ultimate common ancestry, but also that the form of any creature is not solely a reflection of the structure of its individual genes. Instead, adult anatomy is the endpoint of a developmental process that is heavily influenced not just by the underlying genes themselves, but also by the sequence in which the genes are switched on and off; by exactly when this switching happens; and by how strongly the genes themselves are expressed while they are active. This multilayered process (genes, timing, activity) explains the apparent paradox of extreme genomic conservatism together with huge anatomical variety among organisms. And, at the same time, it limits future possibilities. For while changes in the genetic code occur at an astonishingly high rate as a result of simple copying errors (mutations) when cells multiply, few such changes survive in the gene pool. Some mutated genes may linger simply because they don't get in the way (and they may, indeed, turn out to be useful in the distant future, though that won't count for much at the time); but not many will produce a viable result, let alone an adaptively advantageous one. For all these reasons, radical makeovers of the basic structures of heredity are simply not in the cards.

THE ROLE OF CHANCE

Another big reason for not expecting that evolution should be a process of fine tuning is that not all evolutionary changes are the work of natural selection. Chance—technically known as "genetic drift"—is also a huge factor. As a result of those constant mutations, isolated or semi-isolated local populations of creatures belonging to the same species will always tend to diverge from each other purely as a result of what is known as "sampling error"—even in the absence of significant selective forces for change. This is especially true if those populations are small, because the smaller your sample size, the greater your chances of such error. Just think of flipping coins instead of mutations. If you flip a coin only twice, there is a good chance it will come up heads both times; if you flip it ten or a hundred or a thousand times, it is progressively less likely it will always show heads. Tiny populations are equivalent to just a few flips.

Of course, it's also true that not all mutations are equal. Some will have little or no effect on the adult organism; but a few may have a radical influence on developmental processes, and thus upon the creature's final structure. Also important are differences in the degree to which a gene's effects are expressed, or how active its products are in determining the final physical outcome. For all these reasons we should not expect significant evolutionary change in physical form to happen always, or even usually, in tiny and incremental steps. As we will see, sometimes a very small change in the genome itself can have extensive and ramifying developmental results, producing an anatomical or behavioral gap between highly distinct alternative adult states.

None of this is an optimally efficient way to produce adaptation. But, as the luxuriant branching of the Great Tree of Life amply demonstrates, given enough time it *works*. And it works not only as a general explanation of how life diversified over billions of years, but also as an aid to understanding how the deep cognitive gulf separating humans and all other living organisms was so improbably bridged.

This brings us back to the central subject of this book: the story of how human beings came to be the extraordinary creatures they are—as physical entities, of course, but also as an unprecedented cognitive phenomenon. It was a long and eventful (albeit rapid by evolutionary standards) journey from humanity's humble beginnings as a vulnerable prey species, out in the expanding woodlands of ancient Africa, to the position we now occupy of top predator on Earth. But the major outlines of this dramatic story are now becoming clear. And they fit comfortingly well with our emerging views of the multilevel mechanisms underlying evolutionary change. For it's worth repeating that, remarkable as we may think we are, we are actually the product of a routine biological process.

ONE

ANCIENT ORIGINS

Among the most important influences not only on how ancient creatures evolved, but on their preservation as fossils, has been the geography and topography of the Earth itself. This has been as true for our group as for any other, so it's worth giving a bit of background here. During the Age of Mammals that followed the demise of the dinosaurs some 65 million years ago, much of the African continent was a flattish highland plateau. This slab of the Earth's crust lay over the roiling molten rocks of the Earth's interior like a great thick blanket, trapping the heat below. Heat must rise, and eventually ascending hot rock began to swell the rigid surface above.

Thus began the formation of the great African Rift, the "spine of Africa," that formed as a series of more or less independent but ultimately conjoined areas of uplift known as "domes." These blistered and split apart the continent's surface along a line that started in Syria, proceeded down the Red Sea, then south from Ethiopia through East Africa to Mozambique. The Rift's major feature, the Great East African Rift Valley, formed as a complex chain of sheer-sided depressions when the swelling below cracked the inflexible rock at the surface. As the continent continued to rise with the injection of more hot rock from below, erosion by water and wind began to deposit sediments in the valley floors—sediments that contain an amazingly rich assortment of fossils. As a category, fossils technically include any direct evidence of past life, but the overwhelming majority of them consist of the bones and teeth of dead animals that were luckily—for paleontologists—covered and protected by marine or lake or river sediments before they could be obliterated by scavengers and the elements. And, as fate would have it, the sedimentary rocks of the Rift Valley include the most remarkable fossil record we have, from anywhere in the world, of the long history of mankind and its early relatives.

In eastern Africa, Rift sediments began to be deposited in the Ethiopian Dome about 29 million years ago, and similar deposits mark the initiation of the Kenya Dome only a few million years later, at about 22 million years. This occurred during the period known to geologists as the Miocene epoch, and it happens to have been an exceptionally interesting time in primate evolution, as the fossil record shows. It was what you might call "the golden age of the apes," and it set the stage for the evolution of the human family, which appeared toward its end.

Today's Great Apes—the chimpanzees, bonobos, gorillas, and orangutans—constitute a mere handful of forest species now restricted to tiny areas of Africa and a couple of southeast Asian islands. But the Miocene was the apes' heyday, and over its 18-million-year extent, scientists have named more than 20 genera of extinct apes from sites scattered all around the Old World, though mostly in East Africa. The earliest of these ancient apes are known as "proconsuloids." They scampered along the tops of large branches in the humid forests of the eastern African early Miocene in search of fruit, some 23 to 16 million years ago. Like today's apes, they already lacked tails; but in many ways they were more monkeylike, with less flexible forelimbs than those their descendants eventually acquired.

Around 16 million years ago, African climates seem to have become drier and more seasonal, changing the character of the forests. True monkeys began to flourish in the new habitat, and the proconsuloids themselves yielded to "hominoid" apes that more closely resembled their modern successors. Most notably, the apes of the later Miocene developed mobile arms that they could freely rotate at the shoulder joint, allowing efficient suspension of the body beneath tree branches and imparting all-around greater agility. These early hominoids also typically had molar teeth with thick enamel that were set in robust jaws, allowing them to tackle a broad range of seasonally available forest foods as they began spreading beyond the Afro-Arabian region into Eurasia.

In both Eurasia and Africa, paleontologists have found the remains of several different hominoid genera that date back between about 13 and 9 million years ago. These probably represent the group that gave rise to the first members of our own "hominid" family (or "hominin" subfamily; for most purposes the distinction is merely notional). Most of the genera concerned are known principally from teeth and bits of jaw and cranium; but one of them, the 13-million-year-old *Pierolapithecus,* is well known from a fairly complete skeleton discovered not long ago in Spain. *Pierolapithecus* was clearly a tree climber, but it also showed a host of bony characteristics that suggest it habitually held its body upright. Such a posture—in the trees, at least—may actually have been typical for many hominoids of the time (as it is for orangutans today). However, the skull and teeth of *Pierolapithecus* are different from those of any of the putative early hominids that we'll read about in a moment.

WILL THE EARLIEST HOMINID PLEASE STAND UP?

The earliest representatives of our own group lived at the end of the Miocene and at the beginning of the following Pliocene epoch, between about six and 4.5 million years ago. And they appear just as the arrival of many new open-country mammal genera in the fossil record signals another major climatic change. Oceanic cooling affected rainfall and temperatures on continents worldwide, giving rise in tropical regions to an exaggerated form of seasonality often known as the "monsoon cycle." In Europe this cooling led to the widespread development of temperate grasslands, while in Africa it inaugurated a trend toward the breakup of forest masses and the formation of woodlands into which grasslands intruded locally. This episode of climatic deterioration furnished the larger ecological stage on which the earliest known hominids made their debut.

Before we look at the varied cast of contenders for the title of "most ancient hominid," perhaps we should pause for a moment to consider just what an early hominid *should* look like. How would we recognize the first hominid, the earliest member of the group to which we belong to the exclusion of the apes, if we had it? The question seems straightforward, but the issue has proven to be contentious, especially since members of related lineages—such as our own and that of the chimpanzees—should logically become more similar to each other, and thus harder to distinguish, as they converge back in time toward their common ancestor. But while the characteristics that define modern groups should even in principle lose definition back in the mists of the past, attempts to recognize very early hominids have paradoxically been dominated by the search for the early occurrence of those features that mark out their descendants today.

When the Dutch physician Eugene Dubois discovered the first truly ancient human fossil in Java in 1891, he called his new find *Pithecanthropus erectus* ("upright ape-man"). His choice of species name emphasized the importance he attached to the erect stature of this hominid (indicated by the structure of its thighbone) in determining its human (or at least close-to-human) status. But soon thereafter the emphasis changed, at least temporarily. Modern people are perhaps most remarkable for their large brains; and in the early years of the twentieth century, brain size expansion replaced uprightness as the key criterion for any fossil seriously considered for inclusion in the hominid family. Indeed, its big human braincase (which was matched with an ape jawbone) was the basis for recognizing the famously fraudulent English Piltdown "fossil" as a human ancestor in 1912. The fraud was only officially uncovered some 40 years later, although many scientists were suspicious of it from the start; and as time passed the Piltdown specimens became increasingly ignored, which had the effect of bringing the big-brain criterion into disfavor. In its place came a behavioral yardstick rather than an anatomical one: manual dexterity and the manufacture of stone tools became the key to human status, as the notion of "Man the Toolmaker" took hold.

But this too had its difficulties. Eventually and inevitably, attention refocused on anatomy, and various potentially diagnostic morphological features of hominids were touted. Teeth, which are coated with the toughest biological material and thus preserve particularly well in the fossil record, received particular attention. One dental characteristic that many noticed among potential early hominid fossils was thick molar enamel—although, as we have seen, this indicator of a tough diet is also found widely among Miocene apes. Another hominid dental feature that has perennially attracted attention is the reduction in size of the canine teeth. This occurs in conjunction with the loss of honing of the large upper canine against the front premolar of the lower jaw with which it occludes. Large-bodied male apes typically have fearsome upper canine teeth with razor-sharp back edges—although in small females these teeth can be dainty. But again, a tendency toward canine reduction is not unique to hominids. It is also found in various Miocene apes, most famously the bizarre late-Miocene *Oreopithecus,* an insular form that additionally showed a distinct tendency toward postural uprightness. What is more, the remarkable *Oreopithecus* was recently reported to have had "precision-grip capability"—something else that was once thought unique to tool-making hominids.

Part of the problem of spotting features that are unique to hominids stems from the nature of evolutionary diversification. As we look farther back into hominid history, every feature indicative of modern hominids is likely to become less distinctive—and more reminiscent of its counterparts in members of related lineages. Given this reality, it is hardly realistic to expect that we'll ever find an anatomical "silver bullet" that will by itself tell us infallibly if an ancient fossil is a hominid or not. Every effort to do this has foundered on one technicality or another. Take, for example, the early-twentieth-century attempt of the English anatomist Sir Arthur Keith to set a "cerebral Rubicon" of 750 cubic centimeters (cc) minimum brain volume for membership in the genus *Homo.* Any smaller than this, Keith said, and you didn't belong to the club. This was certainly a convenient and easily measurable criterion; and, at a time when very few hominid fossils were known, perhaps it was even a workable one. But predictably, as the hominid fossil sample increased, problems arose. Brain size is notably variable within populations (modern human brains range in size from about 1,000 to 2,000 cc, with no indication that people with larger brains are necessarily smarter), so that even in principle this standard might have admitted an ancient hominid to our genus while excluding his or her parents or offspring. Accumulating fossil finds predictably forced later authors to lower Keith's figure several times, until it became obvious that the entire "Rubicon" idea was misguided.

Similar objections apply to any touchstone of this kind for membership in the genus *Homo* or the family Hominidae. But the temptation to see matters from the "key criterion" perspective is nevertheless always there. Indeed, in recent years paleoanthropologists have come full circle back to Dubois' view, so that the most notable common factor uniting all currently touted "earliest hominids" is the claim that each had walked bipedally on the ground. This seemingly straightforward standard for membership in our family is particularly attractive given that in the latest Miocene the eastern African forests were beginning to yield to patches of more open territory. This would have obliged at least some ape populations to spend more time on the ground (though extinction was, as always, the easier option for steadfastly arboreal types). Still, if this environmental change forced one ape lineage to stand upright, why not others? Several likely did; but only one of them can have been the hominid progenitor.

A further confounding factor is that all of the known "very early hominid" fossils have been found in contexts indicating thickly wooded habitats, or at least mixed ones. The earliest hominids were thus not obliged to walk upright on the ground by the disappearance of their ancestral habitat. We humans have rather reductionist minds, and are beguiled by clear, straightforward explanations. But where murky Mother Nature is concerned, beware of excessively simple stories.

THE CAST OF CHARACTERS

Until close to the turn of this century, the known hominid fossil record extended back in time to only about three to four million years ago. But a remarkable series of finds has since turned up a variety of contenders for the mantle of "earliest" hominid that are significantly older than this. The oldest of them come from around the time that DNA studies suggest our ancestors parted company with our closest ape relatives, believed to be the chimpanzees and bonobos.

"Toumaï" and Orrorin

The most ancient of the "earliest hominids" on offer today is the close-to-seven-million-year-old species *Sahelanthropus tchadensis,* discovered in 2001 in the central-western African country of Chad (well to the west of the Rift Valley). What has so far been published of this form consists of a badly crushed cranium (informally dubbed "Toumaï"—"hope of life" in the local language) and some partial mandibles. These fossils caused a stir when discovered, because nobody had anticipated an ancestral hominid like this. What was particularly strange about Toumaï was that it combined a small (hence rather apelike) braincase with a large, flattish face that was distinctly unlike the more protruding snouts of younger fossil hominids (or apes, for that matter). Two things caused its describers to classify this form as a hominid: first, the teeth. The molars had moderately thick enamel, the canines were reduced, and there was no lower premolar honing mechanism. So far, so good; but as we've seen, both thick enamel and the reduced canine-premolar complex can be matched outside Hominidae. So the key finding was in the base of the crushed cranium, where the foramen magnum, the large hole through which the spinal cord exits the cranium, appeared to be shifted underneath the skull to face largely downward. This is significant in that you would expect to find this setup in an upright biped like us: a skull balanced atop an erect spine. In a quadrupedal chimpanzee, the skull hangs on the front of a horizontal spine, so the foramen magnum has to be at the rear of the skull, facing backward. Unfortunately, though, the skull of *Sahelanthropus* was badly crushed, so the crucial claim about its foramen magnum was inevitably disputed.

In response, researchers took CT-scans of the crushed skull in a medical scanning machine, and produced a computerized virtual reconstruction that eliminated the distortions. Now, no matter how high-tech the procedure is, there's always an element of human judgment involved in making any reconstruction. But the resulting model of the pristine *Sahelanthropus* skull gave its creators substantial grounds for viewing Toumaï as plausibly—if not definitively—the skull of a biped. There are still some skeptics; but although the bipedality question will never be finally settled until key parts of the body skeleton of *Sahelanthropus* are announced, the reconstruction does appear to give this form the benefit of the doubt.

If Toumaï was a hominid—or even if he wasn't—what can we say about his way of life? Fossils found in the same deposits suggest that *Sahelanthropus* lived in an environment that was well watered, with forest in the close vicinity. This doesn't tell us much directly, but it does say something about the kind of resources that were available to this presumed ancestor. Put this information together with its posture, its habitat, and the general form of its teeth, and it seems reasonable to suggest that *Sahelanthropus* was at least a part-time biped that subsisted on a fairly generalized plant-based diet that would have included fruit, leaves, nuts, seeds, and roots, and probably extended to insects and small vertebrates such as lizards. For the moment it's probably unwise to say too much beyond that, though we'll speculate a bit about such things as the nature of early hominid sociality in a little while.

Almost as old as Toumaï is a form discovered in northern Kenya in 2000 (hence its nickname "Millennium Man"), and technically known as *Orrorin tugenensis*. Found at a number of localities dated to about six million years ago, the materials attributed to *Orrorin* are fragmentary, consisting of some bits of jaw and teeth and several limb bones believed (but not demonstrated) to have belonged to members of the same species. The molar teeth are thick-enameled, squarish, and not too large: all features that might be expected in an early hominid. What's more, an upper canine is encouragingly small. But controversy has centered on the incomplete femora (thigh bones), which are unfortunately broken just where their morphology (anatomical structure) would be crucial for establishing bipedality. Still, what's left is entirely consistent with upright locomotion. At the upper end of the body, a piece of humerus (upper arm bone) has a strong attachment area for a key climbing muscle; and one finger bone is strongly curved. Both of these features are indicative of climbing and branch-grasping, and the fossil animals found in the same area suggest that all had lived in a dryish evergreen forest environment, with a notable absence of grassland ruminants. All in all, the *Orrorin* fossils quite strongly support the notion that bipedal hominids were around in the slowly desiccating eastern African forests some six million years ago—a period during which DNA comparisons among humans and their relatives also suggest it's reasonable to expect early hominids to be found.

"Ardi"

The third entrant in the "earliest hominid" stakes is *Ardipithecus*, a recently ballyhooed primate recovered from rocks in the valley of the

Awash River in northern Ethiopia. In 1994 some fragments attributed to the species *Ardipithecus ramidus* were reported from 4.4-million-year-old deposits at a place called Aramis, from which an almost complete, if rather crushed and distorted, skeleton was published in 2009. Found eroding from the desert rocks in horribly crumbly condition, it had taken scientists over a dozen years to restore and study the skeleton. In addition, an earlier species of the same genus, *Ar. kadabba,* was named in 2001 from fossils found in several nearby localities between 5.2 and 5.8 million years old. Though since slightly augmented, *Ar. kadabba* is represented only by miscellaneous materials from sites scattered in time and space, and their association in the same species is even less secure than in the case of *Orrorin.*

Most *Ar. kadabba* fossils are teeth and bits of jaw. The canines almost rival those of female chimpanzees in size, though they are less pointy, while the molar enamel is uncomfortably thin. Postcranial (below the neck) elements include some small pieces of arm bone, a fragment of clavicle (collarbone) and two finger bones; perhaps most interesting is one toe bone, the youngest of all the *Ar. kadabba* fossils at 5.2 million years. This element is strongly curved (i.e., apelike); but it nonetheless resembles its equivalents in later hominids in how it articulates with the bone behind, a feature taken as evidence for bipedality. The upper limb fossils are said to be much more apelike, arguing for a form that, as typical among early hominids, had a much more primitive structure of the upper body than of the lower part. Associated fossils suggest a wooded environment.

The recent publication of the skeleton of *Ardipithecus ramidus* has given us a uniquely comprehensive glimpse of a putative early hominid. And it is indeed a strange beast. A virtual reconstruction of the badly crushed skull revealed a braincase with a volume of between 300 and 350 cc. This is about the size of a chimpanzee's brain today, and it matched a body equivalent in bulk to a small chimpanzee's, weighing about 110 pounds. Unlike humans, apes have small braincases with large faces protruding at the front. And despite some facial reduction, the "Ardi" skull resembles other presumed early hominids in having essentially apelike skull proportions. Its modestly sized molar teeth have enamel that is reportedly thicker than that in the *Ardipithecus* fragments originally discovered, its canines are much smaller than

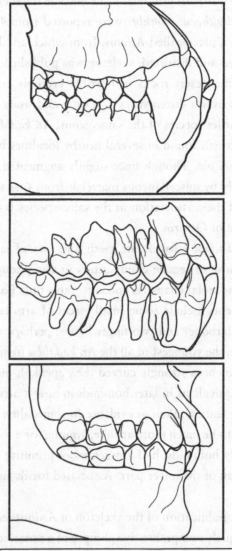

Modern apes (especially males) have very large, pointed upper canine teeth which hone against the front premolars below. In contrast, modern humans have very reduced upper and lower canines that barely project beyond the other teeth, if at all. In determining if a fossil is that of an ape or a hominid, one thing paleoanthropologists look for is evidence of canine reduction. Here we see a side view of the teeth of a presumed male Ardipithecus ramidus *(in the center), compared to a male chimpanzee (above) and a human male (who also, like many of us, lacks wisdom teeth and has an overbite).* Ardipithecus *shows an intermediate condition in which both the upper and lower canines are both reduced, but remain pointy and slightly projecting. Other hominids of the "very early" group show a broadly similar conformation. Illustration by Jennifer Steffey.*

those of *Ar. kadabba,* and there is no premolar honing mechanism. The original Ardi fragments from Aramis had included a piece of cranial base that was said to show a somewhat forward shift and downward orientation of the foramen magnum; and, although incomplete, the new skull reportedly shows the same tendency. All in all, while the reconstructed skull of Ardi does not scream "hominid!" in every respect, its attribution in isolation to an early member of our family might not raise eyebrows excessively.

But what a difference below the neck! The arm and hand bones of Ardi are those of a highly arboreal animal that was well adapted for climbing in trees. Given what we already knew about later hominids, which retain climbing features in the upper body, this came as no surprise. Perhaps more remarkably, though, these bones do not show any of the "knuckle-walking" features seen in the forearms and hands of the chimpanzees and gorillas usually reckoned to be our closest living relatives. Both extant African apes are essentially arboreal creatures (except for adult male gorillas, which are just too heavy to clamber around in most trees). When on the ground, they occasionally rear up and walk short distances on their hind limbs, making displays or even carrying objects; but all apes are basically quadrupeds while on the forest floor— and the long, slender fingers that they depend upon for grasping tree limbs would get in the way during movement on the ground, except for one thing. So when walking on all fours, both chimpanzees and gorillas curl their fingers up into a fist, bearing the weight of the front of their bodies on the outside of the first knuckles. In this way they reduce the effective length of their arms relative to their legs, and this permits more comfortable four-legged walking while also getting those vulnerable long fingers out of harm's way. This unusual accommodation to compressive weight-bearing, by extremities that are basically adapted to the tensile strains of arboreal life, is clearly reflected in the structure of the apes' hands and wrists.

But of course, apes are apes and humans are humans. So why is the absence of any hint of knuckle-walking in Ardi a worry? After all, we *Homo sapiens* show no structural signs of being descended from a knuckle-walking ancestor. The question arises because the molecular systematists who compare the structure of human and ape DNA agree in concluding that

humans are more closely related to chimpanzees than they are to gorillas, sharing more DNA similarities. They are even prepared to hazard
estimates of when gorillas split from the human/chimpanzee group, and
when humans split from chimpanzees, based on the assumption of a more
or less regular rate of change in the DNA molecule over time.

Such molecular age-of-split estimates usually tend to look a little
low to paleontologists: most are in the range of 5 to 7 million years ago
for humans and chimpanzees, with the gorillas peeling off a couple of

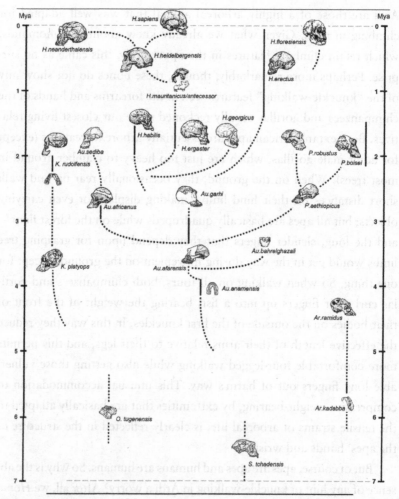

*Tentative hominid family tree, sketching in some possible relationships among
species and showing how multiple hominid species have typically coexisted—
until the appearance of* Homo sapiens. *Diagram by Jennifer Steffey,* ©*Ian
Tattersall.*

million years earlier. But whatever the exact times of divergence, this all means that if the common ancestor of the knuckle-walking chimpanzees and gorillas also walked that way, then so must the chimpanzee-human ancestor. In which case, knuckle-walking must have been lost in the human lineage after the chimpanzee-human split—and you might expect to find some telltale signs of a knuckle-walking past in the wrist and hand of an alleged early human ancestor such as Ardi. The absence of any such signs in Ardi makes you wonder a bit either about Ardi itself, or about our current received wisdom concerning relationships among humans and their closest living relatives.

This mystery isn't going away any time soon. Meanwhile, though, Ardi's discoverers were at pains to emphasize that their fossil's forelimb did not resemble that of either African ape—something that nobody really expected anyway. What is a lot more remarkable is that the rest of Ardi's postcranial skeleton doesn't resemble anything else we know, either. The Ardi pelvis is badly crushed, and had to be restored to its original form using a lot of subjective judgment. As reconstructed, the iliac blades of the pelvis (the elements that flare sideways at the back) are shorter from top to bottom than they are in apes, and are thus marginally more humanlike. What's more, there is a large ridge or "spine" at the front of the pelvis. This structure is associated both with a strong ligament that is helpful in maintaining balance during upright walking, and with a well-developed muscle that helps extend the leg. In humans the ridge is thus quite large, while in the quadrupedal apes it is much smaller. The Ardi team thinks that the short ilia and large spine in their fossil suggest some capacity for upright walking. But in view of the fact that our old late Miocene friend *Oreopithecus* showed these features too, it may be more plausible to associate them with habitual upright posture in the trees than with walking on the ground.

Looking at Ardi's foot reinforces this impression. This is emphatically not what we have come to think of as a hominid foot, where the big toe projects forward in line with the other toes. Rather, it is the long, curving foot of a tree-climber, with a divergent great toe adept at grasping branches. So again, we see a structure in Ardi that is not particularly reminiscent of that of any modern ape; but neither is this foot at all well suited for walking on the ground.

So how *did* Ardi locomote? Right now, that's hard to judge. With a foot ill-fitted for life on the ground, this big-bodied grasping climber weighed so much that its life in the trees would have been hugely restricted had it moved around only on the tops of branches large enough to support its weight. Today the heavy orangutan deals with a similar weight problem by being a "four-handed climber" that frequently suspends itself from clusters of small branches; but the Ardi team categorically denies that its subject shows any anatomical tendencies to a suspensory way of life.

Ardi, then, is a mysterious beast. It has no close living parallels in the structure of its body skeleton, and its cranial construction is at least a little ambiguous. If it is a hominid, it is certainly not directly in the line of later hominids; for not only is it anatomically bizarre but, as we'll see in a moment, there is a much better candidate for the role of hominid progenitor from only a bit later in time. So, if Ardi is a hominid, we have to see it—recent as it may be compared to *Sahelanthropus*—as a late representative of an early branch off the hominid tree. And if that's correct, this strange creature helps, right at the beginning, to set the pattern of remarkable diversity among hominids that was to continue right up to the appearance of our own species. We are alone in the world today; but until very recently there have typically been lots of hominid species around, as the figure on page 12 shows.

WHY BE BIPEDAL?

Ardi forcefully reminds us that the climatically changing world of the Pliocene set the stage for extensive evolutionary experimentation among hominoids, including the exploration of more terrestrial lifestyles. Whatever pressured these creatures to move away from the trees was evidently powerful; for it should never be forgotten that leaving the trees for at least a partly terrestrial life was no small thing. It was, in fact, a huge leap in the dark. In a forest habitat an adept climber, particularly a biggish one such as Ardi, would have been menaced by few predators, at least as an adult. Its food supply would have fluctuated seasonally, but in a relatively predictable way; and its basic lifestyle was underwritten by many tens of millions of years of primate evolution. In contrast, the expanding areas of forest edge, woodland, and grassland would have teemed with ferocious killers such as lions and sabertooths; and at the same time an entirely new foraging strategy would have been required to obtain the unfamiliar resources these habitats offered. For any primate to move into these novel environments meant entering a fundamentally unfamiliar and difficult ecological zone, and for the first hominids it was certainly a huge gamble—albeit one that eventually paid off in spades.

All primates are four-limbed creatures, and why one of them should have taken up erect bipedality on the ground has been incessantly debated. The advantages of this way of getting around are not hugely obvious, while the initial disadvantages—most obviously, the sacrifice of speed in an environment abounding in fleet-footed predators—are manifest. So there really is a big puzzle here. Echoing the default approach to recognizing the earliest hominids, paleoanthropologists have usually framed the "why bipedality?" question in terms of a "key benefit" conferred by this unusual form of locomotion—either in the form of some advantage bestowed by the locomotor style itself, or of some spinoff benefit. Speculations as to what this particular benefit might have been are rife, not least because bipedality opened a host of unique opportunities to hominids.

The ways in which humans have capitalized on those opportunities have caught the attention of paleoanthropologists since the very earliest days of their science. As far back as the mid-nineteenth century, Charles Darwin associated hominid bipedality with the freeing of the hands to modify objects and make tools: a proposal later enlarged by adding the ability to carry things, including food, over long distances. Sadly for the original conjecture at least, it is now known that hominids were bipedal long before they began to make tools.

The array of other speculated advantages to moving upright on the ground is little short of breathtaking in its diversity. At one extreme it has been considered a matter of energetics, and scientists have expended huge efforts to discover how much energy hominoids on the ground use while moving quadrupedally and on two legs. Predictably, the answer is not simple. It all depends on how fast you are going, or on whether you're walking or running, on how rough the terrain is, and on precisely how you're built and move your limbs. In terms of energy used per unit of distance, it's clear that modern humans are more efficient walkers than they are runners; and it has been calculated that on average human running costs are higher than for the average quadruped, while walking costs are lower. So as long as they moved slowly, and avoided the notice of those predators, maybe early hominids saved energy by tottering around on two legs.

But although some researchers have concluded that human bipedal walking is significantly more energy efficient than the locomotion of a quadrupedally ambling chimpanzee, others have been altogether unimpressed by the energy efficiency of modern humans in general. And for an even less efficient bipedal early hominid, costs would have been higher all around than they are for us. This debate will certainly continue, but right now it looks really unlikely that early hominids chose upright walking because it was a more economical way of getting from here to there over open ground.

If you're looking for a physiological explanation for uprightness, a more plausible one is provided by the regulation of body temperature.

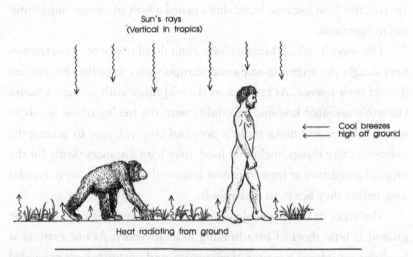

Sun's rays
(Vertical in tropics)

Cool breezes
high off ground

Heat radiating from ground

Some of the consequences of being a biped vs. a quadruped in the unshaded tropical savanna. Compared to the quadrupedal ape, the upright human reduces the area of his body receiving heat from the sun and from the ground, while maximizing the skin area able to radiate body heat. The bulk of the body is also raised off the ground, thus benefiting from the cooling effects of the wind. Illustration by Diana Salles.

Mammals in general need to maintain a reasonably constant body temperature, and the brain in particular is sensitive to overheating. Only a small spike in the brain's temperature can mean irreversible damage. Primates are tropical animals, but they have no special mechanisms for cooling the brain, so the only way for them to achieve this away from the shade of the trees is to keep the entire body cool. If a quadruped stands up out there in the open, the area of its body exposed directly to the hot vertical midday rays of the sun is reduced; and this minimizes the absorption of heat, an important consideration in any animal's temperature budget. In addition, most of the body surface is raised away from the hot ground, maximizing exposure to cooling breezes. This is important for us, because in hot climates humans depend on losing excess heat by the evaporation of sweat. This is a powerful reason, by the way, for believing that the adoption of an upright stance might (at some point) also have been associated with the reduction of evaporation-impeding body hair that is such a remarkable feature of us "naked apes" today.

These elements all add up to a great story, and they may well have somehow been important individually in the early human drama. But as an explanation for the adoption of bipedality, this beautiful theory is, alas, slain by an inconvenient fact: early hominid fossils are generally associated with forested or at least wooded conditions—indicating that bipedality was adopted well before the shelter of the trees was entirely abandoned.

The same observation, by the way, also disposes of the once-popular notion that hominids originally stood upright to see farther over savanna grasses, and were thus enabled to spot predators in the offing more effectively. When you go to the Serengeti Plains today—for most of us, the quintessential vision of Africa—you cannot help but be overawed by the huge open expanses of grassland, and by those panoramas that seem to continue into infinity under the cotton-clouded blue skies. But back in the Pliocene, habitats were typically more closed, and Serengeti-style savannas were pretty much a thing of the distant future. In view of this, some paleoanthropologists have suggested that standing up allows one to reach higher to pick low-hanging fruit from savanna trees—as open-country-living chimps have been observed to do—and have cited this as a possible incentive for early hominids to move around upright. But then, because quadrupedal chimpanzees can do this too, it's obvious that you don't have to be a full-time biped to take advantage of the facultative ability to stand up.

Still, possible benefits of uprightness don't stop with physiology and being effectively taller (also a possible deterrent to predators). Walking upright has been correlated with certain forms of social behavior. One recent suggestion, harking back in some respects to Darwin's original observation, implicates monogamy. By this reckoning, bipedal early hominid males were able to range far and wide for food that they were then able to carry back to their mates, who were tethered to local areas by the burden of their mutual offspring (though it's also been argued that bipedality made it easier to carry infants around). Male bipedality allowed for genital displays to keep the females attracted; while at the same time, by hiding their genitalia between the thighs, female bipedality concealed ovulation so that males needed to be continuously attentive to their mates, reinforcing their fidelity. Well, maybe; but among monogamous primates the two sexes are typically similar in body size, while there is good reason to believe that early hominid females were significantly smaller than males.

The list of potential key advantages (and of objections to them) could go on; but to elongate it here would be to miss the point, for the most important thing to bear in mind when you're wondering why hominids first stood upright is that, once you have adopted bipedality, *all* of its potential advantages are there—and all of its disadvantages too. So perhaps we should abandon the idea of key benefits, and return to the underlying question of why any early hominid would ever have stood up in response to the undoubted challenges of living on the ground—whatever those challenges may have been. And the only plausible answer to this question is that the first hominids to spend any significant amount of time on the ground were *already* most comfortable standing and moving upright. It's clear that the ancestral hominid would never have adopted this difficult terrestrial posture, with all its attendant problems of balance and weight transmission, unless it was simply the natural thing for it to do. Yes, those cute meerkats you see on TV "stand" upright when scanning for predators, but if they see one, they rapidly drop to all fours and scamper away; and this goes for monkeys and living apes as well.

No committed quadruped would ever have walked upright against its instincts purely because of some potential benefit that modern researchers might think up.

Almost certainly, then, the progenitors of our family felt most comfortable lurching around vulnerably on two legs because they were already posturally upright. Presumably they descended from a hominoid ape lineage that habitually held the trunk erect when moving around in the trees—just as the remotely related *Pierolapithecus* and *Oreopithecus* had evidently done as well. This posture would certainly have made sense for creatures that were pretty heavy for tree-dwellers: such animals would have benefited disproportionately from the ability to suspend themselves efficiently by the arms in the small peripheral branches of the trees, where most of the fruit is. Today's African apes are knuckle-walkers because their ancestors were basically arboreal quadrupeds: too committed anatomically to a horizontal stance for their descendants to move upright for any distance on the forest floor, or when venturing beyond the trees. For the early hominids, the reality must have been the opposite: that moving quadrupedally on the ground felt awkward. This is certainly the case for the sifakas of Madagascar, long-legged primates that cling and leap vertically in the trees, and bound bipedally on their rare excursions to the ground

As large-bodied climbers, then, it makes a lot of sense that the hominid precursors should have held their trunks upright when moving and foraging around in the trees. Suspensory orangutans, which tend to hold their bodies erect in the trees, are actually pretty good bipeds on the ground, so perhaps it's legitimate to imagine our remote ancestors, at least in their body form, as "orangutans-but-more-so." Whatever the case, though, the transition from tree-dweller to part-time terrestrial biped must have been difficult, since a climber will find a grasping foot a hindrance on the ground. Most likely, the hominid ancestor lost that type of foot posthaste once it ventured to the ground. But exactly how and in what precise context the in-line terrestrial foot was acquired remains tantalizingly obscure. This deficit in our knowledge is hugely unfortunate because, given that everything that happened later was dependent on the fateful transition from the trees to the forest floor, it presents us with one of the most fundamental mysteries in all of paleoanthropology.

BIPEDAL APES

Not much more than a decade ago, the earliest known hominid fossils belonged to the genus *Australopithecus* ("southern ape"). The first member of this genus was discovered at a South African site in 1924, and numerous others have since been published from localities both there and in eastern Africa (with one central-west African outlier in Chad). But until 1995 all of these "australopiths" dated from between about two and less than four million years ago. Then a new species of *Australopithecus, A. anamensis,* was reported from a couple of sites near the shores of Lake Turkana, a large body of water in arid northern Kenya. The species name given to this form came from the local word for lake (*anam*); and the sediments in which the fossils were found dated from 3.9 and 4.2 million years ago. This extends the range of *Australopithecus* well back in time; indeed, marginally into the "earliest hominid" range of the forms we've just been discussing.

Knowing just how old the fossil-bearing rocks in the Lake Turkana basin are is facilitated by very active volcanism in the region over the past several million years. This is because volcanic rocks contain minerals that incorporate unstable (radioactive) forms of various elements, and these decay to stable states at known and steady rates. When the volcanic rocks—which come in the form of both lava flows and layers of ash-fall that interrupt and interleave the layer cake of accumulating sediments—start to cool after being deposited atop the sediment pile, they do not contain any of the stable products of decay. As a result, any such products that you measure in them must have formed by decay, in a span of time that you can calculate from the known decay rate. Hence you know the age of the volcanic layer, and any fossil-containing sediments lying just above or below it will be (hopefully just a little) younger or older, respectively. Of course, things are rarely quite as uncomplicated as this thumbnail sketch suggests—geological faulting, for example, can tilt, deform, and misalign sedimentary sequences—but over the last half century geochronologists have become quite adept at producing accurate dates, as they have at knowing when the data just aren't good enough to be relied upon. But take note that most of the dates measured in years you'll read about in this book, including all of the early ones, are on rocks, rather than on fossils themselves.

Still, the dating of the Kenyan *Australopithecus anamensis* fossils (and of others some 4.12 million years old from neighboring Ethiopia) is pretty well established. And unlike Toumaï and Ardi, which raise a host of questions despite being represented by more complete specimens, these fossils bear a reassuring similarity to their presumed descendants in the genus *Australopithecus*. What is more, *Australopithecus anamensis* is the earliest hominid we know of that had, beyond a shadow of doubt, acquired important specializations for upright bipedality.

Most of the known fossils of this species are teeth and bits of jaws, but there are some postcranial bones as well, and a particularly vital clue is provided by a broken tibia (lower leg bone). The distal (ankle) end of this bone is especially interesting: it has a large joint surface oriented so as to suggest that the weight of the body was passed directly downward to the ankle joint from the knee, rather than at an angle, as in the apes. This is important because, while apes are capable of moving bipedally, they do not walk upright exactly as we do. Their femora descend directly from the hip joint to the knee in a straight line that is continued downward by the tibia. This is natural in a quadruped, which needs to support itself more or less as a table is supported at each corner by its four legs. But when that same quadruped rears up and walks on two feet, all the rules of balance change. Those two feet are wide apart, meaning that during forward motion each foot has to pivot around the other, describing a wide circle, just as the moving point of a pair of compasses does around the stationary point. This is not only ungainly, but it's hugely wasteful of energy, and apes soon tire when walking upright for any distance. In modern humans, in contrast, each femur angles sharply inward from the hip joint, so that the shaft of the bone forms a "carrying angle" with the vertical tibia below. As a result our knees pass close together when we walk and our feet move in a straight line ahead of us, so that our body weight is not rocked inefficiently from side to side with every step.

The lower leg bone of *Australopithecus anamensis*, which was also reinforced at its knee end like ours, shows that this early hominid had acquired at least the basic prerequisites of efficient bipedal motion. In the upper limb, one wrist bone from Kenya also suggests that this structure was stiffer than that of apes, and more closely resembled the wrist of later hominids. In contrast, while the teeth of *A. anamensis* show general similarities to those of more recent *Australopithecus,* particularly in their thick enamel, in the large, broad premolars and molars, and in the lack of a premolar honing mechanism, in some respects they hark back to earlier times. Thus the incisor teeth are large, even if they are not quite in the league of the fruit-eating apes; the front lower premolar is pointy; and the tooth-rows are long and parallel. The lower jawline also retreats sharply from top to bottom, as in apes. Still, all in all, *A. anamensis* makes a pretty plausible primitive antecedent form for later *Australopithecus,* and it convincingly sidelines the only slightly earlier *Ardipithecus ramidus* as a potential direct ancestor of later hominids.

Associated fossils suggest that *Australopithecus anamensis* typically lived in forest-to-bushland habitats near water, reinforcing the view that the initial adoption of bipedality was not achieved as part of a process of accommodation to encroaching grasslands. Indeed, even after early hominids had developed some important elements of the requisite anatomy, they did not preferentially seek open environments. To complete or at least to augment the picture, a finger bone from Ethiopia is elongated and strongly curved, signifying a powerful grasping hand; and it is believed that *A. anamensis* included agile climbing as part of its behavioral repertoire.

When you place all this in the context of the larger environmental picture, it makes a lot of sense. At maybe 110 to 120 pounds, the average *A. anamensis* probably weighed a little bit more than a typical *Ardipithecus;* but while this is large for a tree-dweller, certainly enough to significantly reduce fears of predation in the arboreal milieu, these primates would have made a tasty morsel for the fearsome predators that prowled the woodlands. So it's probable not only that *A. anamensis* would have sought much of its sustenance in the trees, but that its members would have routinely sought shelter in the branches at night, the time when they were at their most vulnerable.

Although australopiths from later in time are clearly members of the hominid family, paleoanthropologists often like to describe them as

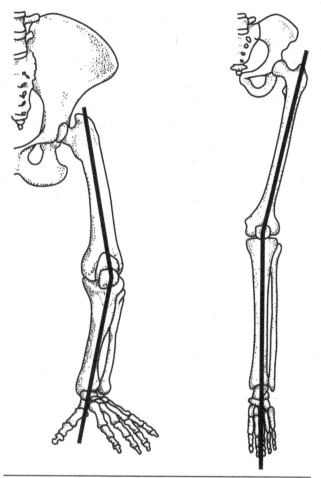

The skeleton of the human leg has accommodated to bipedality in many ways. One of the most important of these is the "carrying angle" between the shafts of the bones comprising the upper and lower leg. The thighbone slants inwards at an angle toward the knee, but the weight of the body is then borne straight downward through the shin bone and ankle to the foot. Because of this geometry, the feet pass close together during walking and running, without requiring the body's center of gravity to move from side to side as weight is shifted from one foot to the other. There is also no sideways component to weight transmission through the ankle. The apes are quadrupeds, and lack these diagnostic modifications for bipedality. In this diagram a modern human left leg skeleton (on the right) is contrasted with its counterpart in a gorilla. The angle at the knee in both primates is emphasized by the intersection of the bold lines. Note also the very different proportions of the pelvis, and the relative length of the legs. Bear in mind that the gorilla is shown in a posture for which it is not specialized (but that the line of weight transmission is basically straight down from the hip joint and somewhat across the ankle), and that the drawings are not to scale: the human leg is in reality much longer than the gorilla's. Drawing by Jennifer Steffey.

"bipedal apes." One reason for this is that australopiths combined humanlike specializations in the pelvis and legs for upright walking with apelike proportions of the skull. They had large, projecting faces hafted in front of tiny braincases, exactly the reverse of what you find in our own skulls, which have tiny faces tucked under the front of the huge globular vault containing the large brain. Another reason is that these early hominids retained characteristics of the forelimbs and torso that would have greatly aided in clambering around in trees. Bipedal they were; but in other respects they were far more apelike than human. Nothing we know of the earlier *Australopithecus anamensis* impedes applying the same description to this form, which is easily fitted into the beginning of the hominid story. Indeed, some authorities point to evidence suggesting that *A. anamensis* was gradually transformed into the successor species *Australopithecus afarensis,* which we will meet in a moment. They are braver than I am; but it is certainly fair to say that in *A. anamensis* we appear to have a worthy precursor for the best-known of all the australopiths, *A. afarensis.*

TWO

THE RISE OF THE
BIPEDAL APES

One minor irony of paleoanthropology is that the major components of the human fossil record were discovered in an order exactly inverse to their geological ages. Our recent relatives the Neanderthals initially came to light back in the mid-nineteenth century, when antiquarianism was still the domain of amateurs; the older species *Homo erectus* showed up a half century later, as the result of the very first deliberate search for ancient hominids in the tropical zone; and the yet more ancient australopiths only became decently documented another half century after that, more or less announcing the dawn of the modern age of paleoanthropology. As a result of this history, the Holy Grail of paleoanthropologists has become the extension of the hominid record into the past.

It's interesting to speculate how differently we might interpret hominid evolutionary history today had the older fossils been discovered first; but while there is no way to know exactly how our views would have differed in that event, what *is* beyond doubt is that the order of discovery of our fossil relatives has deeply influenced their interpretation. Still, the core of this book is a chronological account of the long and astonishing process whereby our ancient ancestor, an unusual but not particularly extraordinary primate variation, became transformed into the amazing and unprecedented creature that *Homo sapiens* is today. And, since trying to interpolate the history of discovery and ideas in paleoanthropology would inevitably have interrupted the flow of the story, I have tried to avoid it wherever possible. But it should never be forgotten that everything we believe today is conditioned in some important way by what we thought yesterday; and some current controversies are caused, or at least stoked, by a reluctance to abandon received ideas that may well have outlived their usefulness. In such cases there will be no getting around a bit of explanation of how we got to our current perspective; and the australopiths are no exception.

THE LUCY SHOW

In keeping with the pattern of paleoanthropological discovery I've just outlined, the geologically oldest hominid species known before the recent spate of "earliest hominid" finds was also the most lately discovered. This is the aforementioned *Australopithecus afarensis,* and its most famous representative is the fabled "Lucy." Lucy was discovered at Hadar, northeastern Ethiopia, in 1974, and she consists of a relatively complete (about 40 percent) skeleton of a tiny hominid individual, usually considered female by dint of her small size. She lived some 3.18 million years ago in a region that is today arid desert, one of the most hostile areas in which humans currently live, but which was much friendlier to hominids back then. The pile of sediments in the Hadar region contains rocks and fossils deposited between about 2.9 and 3.4 million years ago in the valley of the broad, meandering ancestor of today's Awash River. Careful studies of fossils and ancient soils here show clearly that over this period there was some climatic fluctuation, both from drier to wetter and from cooler to warmer. But the area remained one of grassy woodlands overall, with denser forest near the river itself. Sometimes there was more bushland, sometimes less, but trees never grew too far away, and the structure of Lucy's body reflects this.

Lucy's discovery was presaged by the discovery the year before, also at Hadar, of both parts of a hominid knee joint that clearly showed the telltale "carrying angle" between the femur above and the tibia below. Whoever this knee joint had belonged to, there was no question that the knees had passed close to each other during walking, and that the feet had swung straight ahead with each stride. At the time, this was the earliest known evidence of a bipedal hominid by several hundred thousand years. So imagine the excitement and anticipation when the paleontologists went into the field at Hadar the next year, and the too-good-to-be-true feeling when the entire skeleton of a similar individual was unearthed.

Paleontologists don't usually expect to find whole, or even partial, fossil skeletons of land-dwelling vertebrates—too much can happen

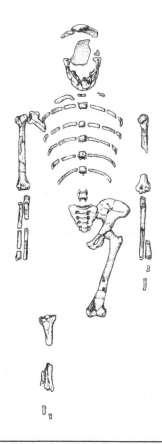

The "Lucy" skeleton, NME AL 188, from Hadar, in Ethiopia. When it was found in 1974 this was the most complete early hominid skeleton ever recovered, and it inaugurated an era of spectacular paleoanthropological discoveries in Ethiopia. Drawing by Diana Salles.

between the moment when an individual dies on the landscape and whenever, if ever, what is left of it becomes buried by sediments. Only a tiny fraction of remains buried in this way are ever again exposed at the Earth's surface by erosion, and then picked up by human collectors before wind and weather have obliterated them. So a tolerably complete skeleton from this incredibly remote period in time was an almost unimaginable piece of luck. In the 1970s, even partial hominid skeletons were virtually unknown before the rather recent era of our close relatives the Neanderthals, who for the first time had hit on the idea of protectively burying their dead. Small wonder, then, that Lucy turned out not to have a complete knee. But the top and bottom elements of the knee *were* preserved on different sides, and they showed the same features as the 1974 knee joint. Lucy had walked upright.

And that wasn't all. In life Lucy had stood not a whole lot more than three feet tall, and had weighed perhaps 60 pounds. (*Australopithecus afarensis* males would have stood up to a foot taller, and would have weighed considerably more.) If you happened by some miracle to meet the diminutive Lucy, you would hardly have recognized her as a particularly close member of the family. But a lot more than her knees attests to her bipedality. The structure that has attracted the most attention in this respect is her pelvis, of which enough remains to make a good reconstruction of the whole. Living apes have narrow pelvises with tall, slender, forwardly sloping iliac blades. The three gluteal muscles that attach behind them are concerned principally with extension of the leg and support of the back during sitting. The tall ilia of apes also raise the lower attachment of strong muscles that go up and across the back all the way to the upper arm and are important in powerful climbing. The modern human pelvis, in contrast, is completely reproportioned. Our pelvises have shortened and become more curved, with more backwardly rotated ilia that efficiently distribute the stresses generated by upright posture, and that cup the abdominal contents lying above. The broad iliac blades also shift the two "lesser" gluteal muscles sideways, enabling them to stabilize the pelvis and upper body during bipedal walking while at the same time being overshadowed, in size at least, by the formerly fairly insignificant gluteus maximus muscle. This has become the biggest muscle in our bodies, and it serves the new purpose of stopping the trunk from tip-ping forward at each foot strike.

Human and ape pelvises are thus significantly different in form, each closely expressing a particular way of getting around. Given that she lived much closer in time to the ape-human ancestor than we do, you might expect to find that Lucy's pelvis had a shape somewhere in be-tween that of an ape and a modern human—perhaps something similar to the reconstructed pelvis of *Ardipithecus*. Amazingly, though, it isn't this way in the least: the *Australopithecus afarensis* pelvis is the very an-tithesis of the high, narrow pelvis of an ape. Like ours, Lucy's hipbones are really short from top to bottom, revealing that the musculature they bore had been reorganized in very much the way that ours has. But her iliac blades are even broader than ours are, showing a dramatic sideways flare. Early interpretations of this unusual anatomy led to the notion that Lucy was a sort of "super-biped," whose pelvis-stabilizing muscles had even better mechanical advantage in bipedal movement than ours do. This breadth and presumed advantage would have been yet further exaggerated by the structure of the ball-and-socket hip joints, in which the head of the femur (the "ball"), which fits into a socket at the side of the pelvis, is connected to the bone's shaft by a "neck" much longer than its equivalent in ourselves.

It always seemed a bit odd that an ancestral biped should have been better adapted than its presumed descendant to the unique upright loco-motor style of hominids. But this strange situation can be explained in terms of the dual function of the pelvis, which provides the birth outlet in addition to gut support and muscle attachment areas. Modern hu-mans face a substantial obstacle in getting a newborn's huge round head through the birth canal, which is why obstetrical problems are relatively frequent in our species. When you look down on Lucy's flaring pelvis, its outline is that of an elongated oval—and the birth canal inside it is oval as well.

Since hominid brains were very small back in Lucy's day, it is thought that such anatomical modification of the outlet in the interests of loco-motor efficiency would have posed no problems for females during the passage of the infant through the birth canal (though a rotation of the baby on the way out might have been necessary). However, it turns out that having a wide birth canal itself has biomechanical consequences, since it affects the spacing of the hip joints. When a biped walks, its pelvis rotates horizontally as each foot swings forward, and this effect is exaggerated the farther apart the hips are, bringing with it a whole slew of biomechanical disadvantages. One reason why human females tend to run more slowly than males is the greater average width of their hips.

While numerous features of the pelvis attest beyond doubt that Lucy was a biped, others reveal that she was not bipedal in quite the same way we are. A similar conclusion emerges from looking at her leg bones, in which that telltale carrying angle at the knee, and a quite convincingly bipedal (though quite mobile) ankle joint, are combined with the remarkably short length of the limbs themselves. Compared to her torso and forelimbs, Lucy had pretty short legs—indeed, her legs were as short as a bonobo's. These proportions were hardly ideal for a strider, but they were a distinct advantage in climbing—and the later lengthening of the hominid leg is widely recognized as a clear sign of a greater commitment to the ground than we see in Lucy. And for biomechanical reasons, such lengthening also permitted some narrowing of the later hominid pelvis relative to Lucy's.

What's more, while Lucy herself possesses only a couple of preserved foot bones, parts of the foot attributed to other individuals of her species indicate that her feet would have been quite long, and her toes a bit curved (although the mid-foot may have been relatively advanced). This was certainly not a committed branch-grasping foot like the ones we see in the modern apes and *Ardipithecus,* with their long, curved digits and widely divergent big toes; but it *is* a foot that would have been substantially more capable in the trees than ours are. The bones of Lucy's upper limb continue the arboreal theme, although relative to the rest of her body her arms were shorter than those of bonobos. Her rib cage, however, tapered sharply upward from its broad base, so that the somewhat upwardly oriented shoulder joints were quite closely spaced. Both of these attributes would have been pretty useful in the trees. And while the Lucy skeleton is a bit short on hand bones as well as of those of the foot, hand elements from other *Australopithecus afarensis* individuals found at Hadar are much shorter than those of apes, but still show some ape-like features of the wrist bones in combination with finger bones that are

quite curved. They also show markings for strong flexor tendons, signifying a strong grasping ability. Taking everything into account, a picture
emerges of *Australopithecus afarensis* as a creature less fully adapted to
bipedality than we are, but much more capable than us in the trees.

This is a configuration unlike anything else we know, except
among other australopiths. Certainly as far as their locomotor and
habitat preferences go, it would be misleading to think of Lucy and
her companions either as an advanced form of ape, or as a primitive
form of human. Her species, and its similarly proportioned relatives,
had hit upon a unique solution to the challenges of living and moving
in the new environments presented to them by climate change and the
fragmentation of the forests.

But Lucy and her like are certainly not frequently (if inaccurately)
described as "bipedal apes" purely because of their odd combination
of bodily features. In the structure of their skulls they show a similarly
unprecedented amalgam of characteristics. Lucy herself has only a lower
jaw and some tiny fragments of the cranium. But as two 3-million-year-
old crania also discovered at Hadar eloquently attest, the general skull
proportions of *Australopithecus afarensis* are broadly apelike, in the
sense that they combine a small braincase (which had contained a brain
not much bigger than that of an ape of similar body size) with a large
and forwardly projecting face. That face, however, has very robust jaws
that house teeth distinctly different from those of any ape. In the upper
jaw, the central incisors of *A. afarensis* are large and flanked by considerably smaller teeth, much as you see in the African apes; but immediately behind the incisors the aspect of the tooth-row changes. As among
the "very early hominid" contenders described in chapter 1, the canine
teeth are reduced in size, even if they are not exactly dainty. The honing mechanism against the lower front premolar is also essentially gone,
although traces of it remain. The rear premolar teeth are broad, and the
molars behind them are flattish and quite large relative to the jaw, setting
the pattern of "postcanine megadonty" (big chewing teeth) that was to
characterize early hominids for some time to come.

The presence of these large molar teeth means that the tooth-rows of
A. afarensis are long, as those of apes are. But largely because the canine
teeth are small, the dental arcades are slightly curved in outline and lack

the aggressively parallel arrangement exhibited by the apes. The dentition of A. *afarensis* is thus rather like the cranium itself in possessing features reminiscent both of apes and of later hominids.

A recent sophisticated study of the wear produced by chewing on some A. *afarensis* molar teeth suggested that, while members of this species probably preferentially sought out soft fleshy fruits to eat when obtainable, where such foods were unavailable they would have gone after tough, brittle foodstuffs like nuts, seeds, roots, and the underground runners of grasses. This would have made them considerably more omnivorous than today's apes are, and it would be consistent with the typically quite heavy wear that we see overall on A. *afarensis* teeth. A diet of this kind suggests a pretty generalist adaptation to a habitat that spanned the spectrum from closed forest to open woodlands.

The Hadar desert badlands have produced an incredible quantity of hominid fossils over the years, and many localities there have yielded fossils of *Australopithecus afarensis*. Undoubtedly the most extraordinary of these localities is a spot known as AL 333 at which, in 1975 and later, researchers unearthed a trove of some 240 fossils representing the remains of 17 hominid individuals—and, most unusually, fossils of very little else. How these bones came to be buried where they are is a puzzle. They are all broken up, which is consistent with their having been transported from somewhere else by water. But why were they concentrated in one place? They weren't accumulated by a scavenging agent such as hyenas (which are famous for transporting hominid cadavers to their dens) because, although broken, they show no signs of gnawing—and a sluggish river channel is, in any case, hardly the spot you'd expect to find a hyena den. So there is a bit of a mystery here; and it's important that it be solved eventually, because despite the fact that the fine-grained sediments in which the bones were enclosed are typical of those laid down in a slow-moving river, one suggestion is that these are the remains of an entire unfortunate social group that was swept up in a single catastrophic event—maybe a flash-flood—that happened at some time between 3.18 and 3.22 million years ago. And if all of the individuals—nine adults, three adolescents, and five juveniles—actually belonged to one social unit, then all of them must have belonged to the same species.

This is not otherwise a sure bet because, although all the comparable fossil parts from Site 333 look basically the same, the size range among them is huge. Still, despite all the uncertainty surrounding how the fossils came to be jumbled together in one spot, the current majority opinion is still that all of the Hadar hominids—including Lucy, who is as small as the smallest of the 333 specimens—belonged to the single species *A. afarensis,* which must consequently have varied greatly in size. The most plausible explanation for the large size range among members of the same species is that males were very much larger than females, comparable to what we see among gorillas today, and not at all like chimpanzees and bonobos, in which sexes differ much less in size.

LAETOLI

Around the time the first Hadar discoveries were being made, another group of paleontologists was hard at work at the site of Laetoli, a thousand miles to the south. Laid down in the Tanzanian portion of the Rift Valley near the well-known site of Olduvai Gorge, the geological layers at Laetoli are slightly older than those at Hadar, running from about 3.5 to 3.8 million years old. Between 1974 and 1979 the broken jaws and teeth of three hominid individuals were collected at various Laetoli localities, but the site is most famous for the numerous animal trackways discovered there beginning in 1976. These include footprint trails left by hominids who, some 3.6 million years ago, had walked across a layer of wet cement-like volcanic ash that subsequently hardened. This was an extraordinary find. We can be confident that Lucy walked upright; but we must always remember that this is not something we can observe directly in the bones. Rather, we have to infer it from Lucy's anatomical structure. A footprint, however, is different, in that it is truly fossilized behavior. And the trackways at Laetoli are as eloquent of bipedality as it's possible to get. At one site an arrow-straight double trail of prints some 80 feet long, more or less like those anyone might leave walking along a wet beach, attests clearly to a purposeful bipedal gait. What is unusual is that the Laetoli environment at the time these prints were made was quite open; the hominids were slogging across a flat plain largely devoid of trees, and they must have felt pretty vulnerable as they did so. But they were heading directly for the Olduvai Basin, only a few miles away, which at that time would have offered all the hospitable resources of a forest surrounding a shallow lake.

The footprints themselves are clear evidence of bipedality: there is no indication that the hominids steadied themselves using their forelimbs, and the way in which weight was transmitted from one end of each print to the other seems to reflect the way we walk—which is to say, it went from the heel, along the side of the foot and across the ball, with a final thrust concentrated on the big toe. This was not the lurching gait of a bipedal bonobo. The feet that made the prints were structured essentially like ours, with longitudinal and transverse arches and a short big toe set in line with the others. The short distances between successive footfalls suggests that even the bigger individual was of fairly diminutive stature, although it seems the pair was not moving very fast—hardly surprising, given the slushy surface across which they were making their way.

While there is nothing to cast doubt on the bipedality of these 3.6-million-year-old hominids, there has been some debate about the exact gait they employed. Did they, for example, fully extend the knee with each step? Or did they retain some vestige of the bent-kneed gait that today's apes use when moving upright, and which, at some remove in time, the hominid ancestor must also have employed? A recent experimental study, using human subjects moving both straight- and bent-kneed, has confirmed that if you don't fully extend your knee, the impressions your toes make in wet sand are deeper than those made by your heel. And the Laetoli prints clearly show heel and toe depressions that are about the same depth, arguing for a straightened knee. Clearly, in these footprints we have evidence of a serious biped.

The scientists who carried out the experimental work suggest that adopting upright locomotion on the ground allowed the Laetoli hominids to increase their ranging distances without expending extra energy, during a period when the forest was diminishing. Indeed, it's very unlikely that any hominid could have made a decent living in the rather barren ancient environment adjacent to the trackways, making it all the more plausible that the prints in the wet ash had caught them in the act of aiming straight for the forests that lined the nearby Olduvai Basin.

Just who those bipeds were is another matter. Not far from the foot-print tracks at Laetoli are rocks of about the same age that yielded the handful of hominid fossils already mentioned. In an unusual collabora-tion, the scientists who initially studied the Hadar and Laetoli specimens eventually decided that they were all from the same new sort of hominid. This new species, *Australopithecus afarensis,* was named for the Afar re-gion of Ethiopia in which Hadar is situated, and from which most of the fossils in question came. But under standard zoological procedure, every new species has to be based on a "holotype," a single specimen to which every other individual assigned to that species has to be compared. And to emphasize their conviction of unity, the scientists chose a lower jaw from Laetoli as the holotype of *A. afarensis.* Not everybody found this appropriate, though, as some scientists felt that they could discern evi-dence for more than one species of hominid just at Hadar, let alone at the Ethiopian and Tanzanian sites together.

At present there is a state of uneasy truce, with most paleoanthro-pologists willing to accept at least provisionally that the known bones and teeth, at least, can be assigned to the same species. But the associa-tion of *A. afarensis* and the footprints is much more actively debated. Perhaps a majority of paleoanthropologists is willing to believe that in-dividuals of *A. afarensis* made the Laetoli trails; but at least a substantial minority thinks that the fossil foot bones from Hadar indicate a foot far too long and primitive to have produced the strikingly modern Tanza-nian footprints. If the majority is right, then we will have to accept that the Lucy's arboreal adaptations, and her broad pelvic proportions, were indeed compatible with remarkably humanlike bipedality. But the jury is still out; and all we know for sure right now is that *somebody* was out there strolling upright through the Tanzanian Rift 3.6 million years ago.

DIKIKA

Only a few years ago, the word "Dikika" was in almost nobody's vocab-ulary, but now it's one of the hottest buzzwords in paleoanthropology. During the glory days at Hadar in the 1970s, and again in the 1990s, ev-erybody was too busy to look south across the Awash River to the more or less equivalent deposits of Dikika. But when the investigation of those rocks finally began at the start of this century, they proved to have a dramatic story to tell. First, some scrappy bits of tooth and jaw attributed to *Australopithecus afarensis* showed up; but these were soon overshadowed by the discovery of the crouched partial skeleton of a three-year-old juvenile. Thought to be female, the skeleton was soon baptized with the informal name "Selam" ("peace"). So well preserved was Selam that it seemed the infant was snatched from its group by floodwaters and almost immediately buried whole in soft mud, some 3.3 million years ago. And this poignant Pliocene misfortune proved a bonanza for the paleontologists, who found that the Selam fossil preserved elements that were not included, or were more poorly preserved, in the extensive *A. afarensis* collections from north of the Awash. Among these elements are a hyoid—the bony portion of the Adam's apple—that resembles that of an ape rather than a human, and a complete scapula (shoulder blade) that is unexpectedly reminiscent in overall shape of a gorilla's. Selam has the ankle of a biped, but the carrying angle between her femur and tibia is not marked, confirming that this is a feature that has a strong component of behavior in its development—carrying angles do not properly develop in modern humans who spend their entire lives in wheelchairs.

One big difference between apes and humans is that apes develop to maturity much faster than we do, depriving them of the extended childhood that provides us with so much opportunity to learn. On her own, Selam cannot do much to demonstrate how quickly individuals of *A. afarensis* developed; but the expectation would certainly be that she lay on the apish side of the curve. The fossil has yet to be fully freed from the hard rocky matrix that envelops it, but as far as we can tell, her upper body structure confirms the generally arboreal features found in adult *A. afarensis*. Not only does Selam's scapula show a shoulder joint that was oriented largely upward, as a good climber needs for holding the arms above the head, but her hand also shows features associated with climbing. Ape and human hands may look superficially similar, but they are in fact constructed very differently. Apes have thumbs that are short in relation to the fingers, and the hand is long, with its major axis in line with the arm. The ape hand is more the hand of a powerful grasper than of a dexterous manipulator; and it is the kind of hand that you want if you are going to spend most of your time clambering around in

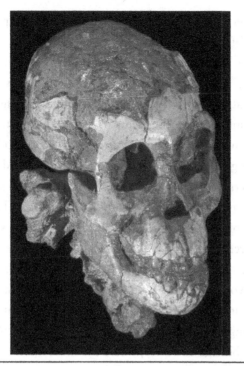

The skull of "Selam," the infant skeleton found at Dikika, Ethiopia. Despite its tender age of only three years, this tiny 3.3 million-year-old skeleton has yielded a wealth of information about the structure and development of the species Australopithecus afarensis. *Courtesy of Zeresenay Alemseged.*

the branches of trees. In contrast, the major axis of the modern human hand goes across the palm, and the thumb is long and can be opposed precisely to any of the other, shortened, digits. Selam's digits appear to have been long and curved.

To complete this image of a hominid that was not completely wedded to the ground, CT scans of Selam's ear region suggest that the semicircular canals of her inner ear resembled those of apes and other australopiths. These canals are important organs of balance, and their orientation reflects not only the way the head is habitually held, but how well it is insulated from movements of the spine on which it is poised. Selam's semicircular canals are reported to resemble those of apes and other early bipeds, suggesting that although her species may have been an upright walker, it was not suited for fast running—an activity in which it is important to maintain a fairly constant head position despite the gyrations of the body below.

But Selam is not the only surprise from Dikika. In mid-2010 the research group there came up with something that was even more remarkable. Strata dated to just under 3.4 million years ago yielded four superficially unimpressive fragments of mammal bone that proved, on close inspection using a scanning electron microscope, to bear markings that archaeologists suggested were of the kind produced only by stone tools. To understand the significance of this finding, you need to bear in mind that 3.4 million years ago is *800 thousand years* before we have any evidence of stone tools themselves. The very first stone tools we know of are reported from not very far away along the Awash Valley, but are a mere 2.6 million years old. Yet stone tools are very durable things—carnivores don't chew on them, and under most circumstances they preserve in the record indefinitely. If ancient hominids at Dikika were clobbering one rock with another to produce sharp cutting flakes so they could butcher carcasses, where are those flakes? And where are the "cores" with the flakes removed from them? It's not as if paleoanthropologists haven't thoroughly scoured the Dikika and Hadar landscapes for interesting objects of this age.

There are several possibilities as to why no stone tools have been found in these regions. One is that the paleontologists had the wrong "search image," and simply were not finding such implements this far back in time. But even the most primitive deliberately made tools show distinctive signs of manufacture, and it's unlikely that over many years experienced searchers would totally miss pieces of stone with obvious modifications. Alternatively, the Dikika researchers suggest that the very early history of stone tool-making showed very low "intensity": i.e., just one flake was removed per core, so that each core would show little sign of modification while the flakes themselves were rare. Another possibility is that the scratches are in fact "trampling marks" produced by the sharp hooves of grazing mammals that had stepped on the bones. But, perhaps most likely, the australopiths had simply used naturally broken stones for butchery. Experimental archaeologists have now shown that it is indeed possible to dismember a mammal carcass using stones of the kind that are routinely fractured against each other while being swept down rivers. Such pieces don't have the razor-sharp edges of deliberately manufactured stone tools, but they can nonetheless do the job.

Still, whatever exactly happened, under the microscope two of these slivers of bone (a piece of rib and a fragment of femur, one from an animal the size of a cow, the other the size of a goat) not only show "cut-marks" of the kind that are produced by slashing with a sharp tool while the bone is still fresh, but they also bear scratches and pits such as those made when a piece of fresh bone is scraped or bashed with a hard and pointy object. So here is a strong suggestion that early hominids, presumably *Australopithecus afarensis,* were indeed butchering the carcasses of large animals out there on the bushy Dikika landscape, some 3.4 million years ago—even if they weren't, strictly speaking, stone toolmakers.

UP THE RIVER

Dikika presumably has plenty more surprises in store for paleoanthropologists. Doubtless so also has an area, some distance upriver, that is known as the Middle Awash Valley. This region occupies a unique position in paleoanthropology. It has not produced fossils as lavishly as Hadar has, but it has yielded hominid remains that range from *Ardipithecus kadabba* at 5.8 million years ago, to the very earliest days of our own species *Homo sapiens* a mere 160 thousand years ago. No other place in the entire world registers events in hominid evolution over such an enormous span of time. The 4.12-million-year-old Ethiopian *Australopithecus anamensis* fossils are from the Middle Awash, and from relatively close by comes a geologically younger partial hominid skeleton assigned to *Australopithecus afarensis,* from a 3.58-million-year-old locality known as Woranso-Mille.

The Woranso-Mille bones are substantially earlier than any of the *A. afarensis* fossils from Hadar, and they are comparable in age to the Laetoli scraps. Unfortunately the skeleton lacks a skull or teeth, but the skeleton is said to be broadly similar in preserved portions to the later and smaller-bodied Lucy. A scapula is quite well preserved, and although its shoulder joint portion does seem to have been quite upwardly oriented, it is unlike its counterpart in the Dikika child in bearing no particular resemblances to its equivalent in any African ape. As for the lower part of the body, the Middle Awash researchers think that that their bigger-bodied specimen is more relevant than the diminutive Lucy for assessing exactly how *A. afarensis* walked, as smaller subjects weigh less and therefore require fewer specializations to support their body weight. This is actually pretty arguable, for one thing because it's hard to imagine a pelvis and leg structure more suggestive of bipedality than Lucy's. But it is certainly good to have a larger counterpart skeleton. The Woranso-Mille individual sadly lacks a complete leg, but by the team's estimate this hind limb had been relatively a bit longer than Lucy's. If this is indeed the case, then the Woranso-Mille fossil might fit a bit better than the Hadar materials with the sort of hominid that made the more or less contemporaneous Laetoli prints. Frustratingly, it has no foot bones, leaving lots of room for speculation on this point.

Moving up the Middle Awash geological section, some jaws from about 3.4 million years ago have also been attributed to *Australopithecus afarensis*. These are interesting, but don't tell us a lot more than we already knew. The picture becomes a lot more exciting about a million years later—which is well after *A. afarensis* had disappeared from the record at Hadar. Fossils from a 2.5 million year-old site called Bouri have been given the name *Australopithecus garhi* ("garhi" apparently meaning "surprise" in the local language). And although the specimens concerned are not that impressive—consisting mainly of some cranial fragments that include the brow and a reasonably complete upper jaw with teeth, plus a few postcranial bones—they certainly did come with a surprise attached. The Middle Awash team claimed that the arm and leg bones, combined with a slightly longer hind limb than Lucy's (as in the as-yet-undiscovered Woranso-Mille skeleton), indicated powerful limbs. And while they did not directly associate the limb bones with the skull parts, the reconstructed skeletal proportions may have been a factor in their claim that they had found a new and "advanced" form of *Australopithecus* that was directly antecedent to our own genus *Homo* (even though the teeth preserved in the palate are biggish, and closely resemble comparable specimens from Hadar). However, the official reason for this conclusion was that the Bouri fossils were "in the right place, at the right time" to play the role of ancestor for *Homo*, regardless of any anatomical particularities that might have tied the Bouri form in with any of its proposed descendants—or with *A. afarensis*.

All of this may sound as if the Middle Awash area has somehow tantalizingly preserved the record of a steady progression of hominids at regular intervals in time. But you need to know that all interpretation of the Middle Awash materials by their discoverers has been conditioned by the underlying belief that the story of human evolution has essentially been a linear one. The idea is that a single central lineage gradually transformed under natural selection from one species into the next, until the primitive *Ardipithecus* had been transmuted into the finely burnished *Homo sapiens*. This is a perspective that gives greater significance to the age of each fossil concerned than to its anatomy. And while there is a certain logic here, it is a logic that only applies if you think of evolution as a steady chain of species running through time. You will have already gathered that this is not the only way of viewing either the evolutionary process itself, or the shape of the human evolutionary drama that resulted from it; but, as we'll see in some detail later, it is a viewpoint that has lingered with particular tenacity among paleoanthropologists.

Whatever the underlying process, it wasn't the Bouri fossils themselves that were the real surprise. An accompanying article described some mammal bones from the Bouri deposits that clearly bore cut-marks made by sharp stone flakes—and remember, this was a decade before the comparable finds downriver at Dikika. During the second half of the twentieth century, the notion of "Man the Toolmaker" had exerted a powerful attraction upon paleoanthropologists. Making stone tools, it was thought, was the key behavior that had sent humans off on their unique path, and was thus the defining attribute of humankind. Very early stone tools, over two million years old, had been reported since the 1970s from sites in Kenya and the Omo Basin of southern Ethiopia; and tools as much as 2.6 million years old were shortly to be announced from the nearby Middle Awash locality of Gona. But at the time, the Bouri cut-marked bones represented the earliest evidence of tool use by any ancient hominid—and the only hominid that could be implicated in this behavior, even though the association was not definitive, was *Australopithecus garhi*—a bipedal ape. Once it had been realized that stone tools were already being made over two million years ago, the hunt had of course been on for early *Homo* in the same time range, and some fossil fragments arguably attributed to early members of our genus had duly turned up. But the anatomically primitive nature of the Bouri find demanded a rethink.

No actual stone tools were found at Bouri, but those from Gona resembled the simple "Oldowan" tools (named for Tanzania's Olduvai Gorge, at which such implements were first identified) that were already known from later sites. These were small cobbles of mainly volcanic fine-grained rock that had been hit with a "hammer" stone to detach one or more sharp flakes that could be used for cutting. Often the cores themselves show signs of being used for pounding, an activity that has been associated with the characteristic torsional (twisted) fractures that occur when long bones are bashed with hard objects to extract their marrow.

Crude as Oldowan utensils might appear, they are remarkably efficient, as archaeologists have shown by butchering entire elephants using flakes an inch or two long made using Oldowan techniques. What's more, the effectiveness of these simple implements is also evident in the fact that stone tool kits barely changed for a million years after Gona times (and almost two million after Dikika), even as new kinds

Partially reassembled replica of an "Oldowan" stone tool, consisting of a pebble of fine-grained volcanic rock with several sharp flakes successively chipped from it. Replica by Peter Jones; photo by Willard Whitson.

of hominids came and went. Clearly, this was a highly successful technology that did everything that was demanded of it.

Whoever it was, exactly, that had made the Gona tools and the Bouri (and Dikika) cut-marks, these extraordinary finds are witness to a revolutionary behavioral innovation among hominids. Extensive coaching of a bonobo called Kanzi—a star in ape "language" experiments, and a cognitively admirable representative of his species—failed to teach him to hit one rock with another at exactly the angle and force necessary to detach a sharp flake. He rapidly got the idea of using such flakes to cut a cord that held a piece of food just out of his reach; but he never really picked up the principles of shaping stones. Eventually he developed a preference for throwing a rock on the floor to shatter it, and then picking through the fragments to find a sharp one. This may actually have had as much to do with Kanzi's hands as his brain and learning capacities. Making stone tools is not only hard on the hands, but it also requires a hand that is capable of holding objects precisely.

Our hands, with their broad palms, long thumbs, and ability to oppose the thumb to the tips of all the other fingers, are ideally structured to manipulate objects. This ability demands the rearrangement of a whole host of palm muscles to promote delicate movement rather than strength. The hands of living apes are, in contrast, very differently proportioned. They are much longer and narrower than human hands, and the muscles and tendons are arranged to flex the long fingers with enormous power—which is exactly what you want when you spend most of your life hanging on to tree branches. What's more, because of the apes' knuckle-walking proclivities—whereby they bear the weight of the front of the body on the outside of the flexed fingers when they're ambling around on the ground—the tendons of the apes' hand flexor (closing) muscles are shorter than those of the straightening extensor muscles, making it impossible to extend both the wrist and fingers at the same time. This kind of strongly flexing hand is far from ideal for making tools, an activity that requires hugely precise movement and placement of the digits.

Exactly why the early hominid toolmakers already possessed hands that were up to this unusual task is not clear. Logically, there must have been some advantage to losing the specialized grasping capacities of the apes, a development that we already see quite strongly expressed in the hand bones from Hadar despite the still-curved fingers. And chances are there were no knuckle-walkers in their ancestry. But whatever the countervailing benefit was, it wasn't the ability to make stone tools, which as far as we can tell only started to be done well after the australopiths and their various physical characteristics, including the manual ones, were already well ensconced. Still, even though we are in the dark as to the exact circumstances, there is a really important lesson to be gained from this episode in human evolution: you can't use a structure until you have it. As a result, most of our so-called "adaptations" actually start life as "exaptations": features that are acquired through random changes in our genetic codes, to be co-opted only later for specific uses. Natural selection is, quite simply, in no position to drive new features into existence, no matter how advantageous in theory those features might be. This is yet another reason why the idea that the Dikika hominids butchered animals using naturally fractured stones is attractive, for it is otherwise very odd that there is no evidence of stone tools or their use in the long lapse of geological time between the Dikika and Bouri/Gona finds.

THREE

EARLY HOMINID LIFESTYLES AND THE INTERIOR WORLD

Emerging from the deep forests into the forest edges and the adjacent woodlands and bushlands was a major commitment for the early hominids. And it was one with multiple consequences. Not only did this ecological shift involve a radical change in diet and locomotion, but it also brought with it a huge increase in vulnerability to predators. What's more, it laid bare several fundamental differences between our ancestors and their close ape relatives. While apes move around on all fours when they venture into savanna environments, the hominids evidently responded to the new challenges not only by standing upright, but also by adopting a radically new dietary strategy. The consequent changes made enormous new demands on their various bodily systems. Let's see how they adjusted.

STAYING WELL FED

A major imponderable in paleoanthropology is how the early hominids contrived to cope with a new diet that evidently included animal fats and proteins. Even today, we meat-eating humans have digestive tracts that more closely resemble those of our vegetarian ancestors than they do those of carnivores. And our teeth, small as they have become, are essentially those of plant-eaters, emphasizing grinding rather than carnivorous cutting. But at some point those ancient vegetarian bipedal apes became interested in animal carcasses: an interest based on the potential of those cadavers to provide food. Dealing with this unfamiliar dietary supplement posed a host of problems.

Red meat straight from the carcass would have been indigestible for those early hominids, whose stomachs would not have been filled with the highly concentrated acids that allow modern carnivores to break down bone and muscle tissues before delivery to their short intestinal tracts. One possibility is that our forebears used those multiply bashed rock cores to pound red meat in order to break down the muscle tissue and render it somewhat more digestible. Another is that they avoided muscle tissues altogether, and concentrated on eating the internal organs of dead animals. We know this happened at least sometimes in the early days of the hominids: a 1.7-million-year-old skeleton from northern Kenya shows distortions of the bones probably caused by an excessive intake of vitamin A, most plausibly from a carnivore's liver. But a specialized interest in offal seems unlikely, not only because these organs provide the preferred foods of the primary carnivores and scavengers with which the hominids must have competed, but because of those telltale cut-marks. Some of those marks, at least, must have been made in the process of removing red meat from the limbs, for they would not have been made where they were in the process of simply opening the abdominal cavity and removing organs.

Another possibility that has received a lot of recent publicity is that hominids made flesh more digestible by cooking it over fires. Cooking certainly makes nutrients in both plant and animal foods much more readily available to the enzymes of the stomach; and there is some evidence that people today find it hard to maintain their weight on diets composed entirely of raw food. The big difficulty here, though, is that there is no good evidence for controlled fire before about 800 thousand years ago—and the regular use of fires for cooking seems to have begun only well after that. Still, some authorities think that the increase in average hominid brain size that began some two million years ago could *only* have been made possible by higher-quality (i.e., higher-fat, higher-protein) diets than plant materials alone could have supplied. The brain is an extremely energy-hungry organ, whose consumption of calories increases with its size. You simply can't maintain a brain that is larger than absolutely necessary without some caloric compensation; so it is argued that some animal protein must have been available in the otherwise pretty non-nutritious hominid diet to fuel the brain's enlargement.

There are various independent indicators that hominids have been eating meat for a long time. Bizarrely enough, one of these comes from the study of tapeworms. Different forms of these ubiquitous intestinal parasites are specific to particular hosts, and it was long thought that humans were first afflicted by them when they domesticated cattle and started living in close proximity to herds. But according to molecular studies, tapeworms got into hominid populations very early on, presumptively as a result of sharing antelope carcasses, and thus remnant saliva, with carnivores—probably lions, wild dogs, or hyenas.

This finding fits with studies of stable isotopes (alternative forms) of carbon preserved in the teeth and bones of australopiths. These studies are based on the principle that "you are what you eat." Most plants fix atmospheric carbon dioxide along what is known as the C_3 pathway. This results in a low abundance of the carbon isotope ^{13}C in the bones and teeth of animals that eat them. Some kinds of vegetation, however, including tropical savanna grasses, use the alternative C_4 pathway. Eating such resources results in a greater quantity of ^{13}C in the tissues of the animals consuming them. The resulting chemical signals, measurable in the teeth, get transferred from animal prey to the predators that eat them. The relative abundance of these isotopes in its tissues thus provides a clue to an animal's diet, irrespective of whether it is or was a primary herbivore or lay higher up the food chain.

Isotope studies have confirmed what was already known from behavioral observations, namely that today all chimpanzees, even ones living in open country, stick to the kind of C_3 diet furnished by the forests. Numerous australopiths, on the other hand, show a strong C_4 signal; and since it is improbable that they were all grazing on grass, this signal must have come from grazers they were eating. Plausible candidate victims include such creatures as hyraxes, or young grazing antelopes.

This does not mean that early hominids were mostly eating meat; but the isotopic signal indicates that they had departed from their ancestral diet of forest plants, and had become significantly more omnivorous. So unless they were raising herds—a practice that began only well after *Homo sapiens* had become its modern self—they had to hunt for their meat or scavenge for it. Among living hominoids chimpanzees occasionally hunt, but they don't significantly scavenge. What's more, while chimpanzees mostly hunt cooperatively, the sharing of their kills seems to be much more significant in reinforcing social bonds within the group than it does in contributing to their diet. And when chimpanzees do hunt, they go after animals—colobus monkeys, blue duiker, bushbabies—that have a forest diet, and thus a C_3 signal in their tissues.

Something different was clearly happening in the case of the australopiths who, wherever they may have spent most of their time, were almost certainly getting most of their C_4 component from the cadavers of grazing animals obtained away from the deep forest. Since they were small bodied and not particularly fleet of foot, the most obvious C_4 source would have been scavenged carcasses; but out there in the open, competition for this relatively rare source of sustenance must have been rough. Even more importantly, dead cadavers quickly turn toxic, and living primates—including humans—lack any of the specializations for dealing with this huge problem that full-time scavenging animals such as vultures have. Once decomposition sets in (which in the tropics will not be very long after the initial hunters have departed), the virus, microbe, and other parasite populations inhabiting the carcass skyrocket, and the flesh rapidly becomes not only indigestible but potentially lethal. Little wonder that scavenging is so rare among living primates: one study of chimpanzees in Uganda found that they spontaneously encountered opportunities to scavenge fresh carcasses about four times a year, but only even tasted the meat about one time in ten, or once every two and a half years. All in all, scavenging of old carcasses does not appear to be a very attractive proposition for primates in general; and why early hominids should have taken to it in any major way—if that's what they did—is not at all apparent.

Still, there *is* that nagging C_4 signal. And one suggestion is that early hominids became meat-*stealers* when they began spending significant amounts of time away from the closed forests. Hominids and leopards,

It was dangerous out there in the woodlands and savannas for small, slow-moving hominids. This artist's reconstruction of a leopard dragging off a juvenile Paranthropus *is based on a braincase fragment from the South African site of Swartkrans which is pierced with holes that exactly match the size and spacing of a leopard's canine teeth. Illustration by Diana Salles after a sketch by Douglas Goode.*

for example, seem to have had a particularly intimate relationship since the ancestral hominid environment was broadened to include woodland and bushland (one australopith skull fragment from South Africa even bears holes made by leopard teeth). Leopards, fearful that larger carnivores will take over their kills, are frequently seen stashing the cadavers of their prey high in trees for safekeeping while they are off patrolling their territories. Australopiths might well have capitalized on their considerable climbing skills to scoot up trees and steal bits of carcass in the leopards' absence, a risky activity that would certainly have been facilitated by the ability to quickly cut off chunks before making a hasty getaway. In that event, perhaps we shouldn't be surprised by discovering that stone tools were first invented not by members of our proud genus *Homo,* but by bipedal apes. This perspective does make it possible to see that the use of stone tools could have made achievable the major dietary shift that underwrote the remarkable developments to come. And at the very least, while this story of stealing fresh meat doesn't come close to tying up all the loose ends, it opens up new possibilities.

WHAT CAN CHIMPANZEES TELL US?

The evidence of tool use, and yet more of tool-making, tells us that the bipedal apes had graduated—perhaps as much as 3.4 million years ago, and at least before 2.6 million years ago—to a cognitive state that lay well beyond anything we can infer for the apes as we know them today. Not only were the early stone tool makers spontaneously indulging in an activity that required an insight into how stone fractures that Kanzi the bonobo simply hasn't demonstrated, but they also showed a degree of foresight that is not reflected in the activities of hunting chimpanzees today. Still, these are evidently not cognitive skills that were possessed by the first hominids to quit the forest; indeed, quite probably at that early stage our forebears didn't even have the biological wherewithal to possess them. Let's also keep this in perspective and remember that chimpanzees are actually very complex beings, as we saw in the opening paragraphs of this book. No human being looking at a chimpanzee will fail to see a lot of him- or herself there, although what the usually caged individual is feeling or experiencing always remains veiled.

On the technological front, the similarities between humans and chimpanzees (and other apes) are keenly reflected in the fact that hardly a month seems to pass in which these primates are not documented to indulge in yet another sophisticated behavior we'd thought only we possessed. The most recently discovered of these behaviors is the spearing on sharp sticks of sleeping bushbabies, though this remarkable practice may well have been superseded as the revelation du jour by the time this book goes to press.

The wide variety of simple technologies that chimpanzees exhibit—and which are passed along through the generations in a form of imitative "cultural" transmission—is at least in part attributable to the range of environments they inhabit. Chimpanzees live in a remarkably wide variety of central and western African habitats ranging from dense rain forests to wooded grasslands. This environmental spectrum resembles that of the early hominids—though a big difference is that, even in drier and more open areas, the chimpanzees tend to select foods—mostly fruits—that resemble the resources available to them in the forests. Tubers and other tough, gritty, open-country foods of the type that the early hominids evidently enjoyed are of little interest to them. Nonetheless, the chimpanzees are highly conscious of the potential resources around them.

At a place called Fongoli, in Senegal, where the environment is a mosaic of trees interspersed with grassland, chimpanzees live alongside a population of bushbabies, small defenseless nocturnal primates that spend their days hidden, often deep inside holes within trees. Evidently the Fongoli chimpanzees regard the bushbabies as tasty snacks, because researchers at the site saw many instances of chimpanzees fashioning wooden "spears" and thrusting them into tree holes, apparently in the hope of impaling a bushbaby. I was relieved to learn that only one such attempt out of 22 recorded was successful, but what is most interesting here—in addition to the unusual fact that not just adult males but also females and juveniles engaged in this kind of hunting—is that the chimpanzees always followed the same, evidently well-established, procedure.

First, a branch was broken from a tree. Then it was stripped of smaller branches and twigs, in much the same manner used by other chimpanzees to prepare the slenderer tools used in "fishing" for termites in their mounds. Often the bark was removed from the proto-spear, and further trimming was done; and on some occasions the spearmaker used his or her incisor teeth to sharpen the spear's working end. Once the 18-inch- to 3-foot-long implement had been completed, it was jabbed forcefully into a tree hole, then withdrawn, inspected, and sniffed. In the single case observed where the hunt was successful, the bushbaby was not withdrawn from its branch cavity impaled on the spear. Rather, after apparently detecting the prey's scent or tissues on the spear point, the adolescent female hunter broke the branch by jumping up and down on it, and then retrieved the inert bushbaby by hand. After this she withdrew and ate the carcass alone. In two other cases, Fongoli chimpanzees were seen eating bushbabies, though it's not known by what method they acquired them. These individuals were also basically alone, though one may have shared her prey with her juvenile daughter.

The use of spears in hunting by chimpanzees is a technological eye-opener, on a par with the amazing recent report that these primates use stone anvils to help in cracking hard-shelled nuts, and the even more astonishing revelation that this behavior has a 4,300-year-old archaeological record in the form of ancient stone scatters. It is also a reminder of the incredible flexibility of chimpanzee behavior, since the Fongoli way of hunting differs hugely from what is seen among chimpanzees in other locales. In more closely forested habitats in both western and eastern Africa, chimpanzees do sometimes hunt alone; but they normally hunt cooperatively, and this behavior is more common than was once thought. Where they live in the same African forests as the widespread red colobus monkey, for example, chimpanzees hunt these primates between four and ten times each month, and their success rates hover above 50 percent. Often hunts are opportunistic, apparently initiated purely by chance encounters between predators and prey; but at other times male chimpanzees seem to actively patrol the forest in search of the colobus. Hunting is largely a male pursuit, and the larger the number of males cooperating in a hunt, the higher the success rate tends to be. The process is something to behold. The tree-living colobus themselves live in large groups that spend their lives in the forest canopy. A hunting party of chimpanzees will surround an entire group, with some of the hunters stationed on the ground and others up in neighboring trees. Some of the chimpanzees actively chase the monkeys, dashing after potential victims with enormous vigor, while others apparently just observe; but all are in a state of high excitement. The chimpanzees have greatest success when they are able to corral one or more fleeing monkeys in a spot where the forest canopy is broken, and they can isolate their victims in a single tree. Once the victim is caught, it is eagerly torn to pieces and shared among those in attendance, each anxious to receive a share.

Some chimpanzee communities may consume hundreds of pounds of monkey meat in a year; but nonetheless, while the monkeys might hence seem to be a valuable dietary resource, more often than not when the two species encounter each other the colobus are ignored rather than pursued, even by chimpanzees out foraging for food. Moreover, if challenged a chimpanzee holding a juicy colobus carcass will more readily give it up than he would a branch full of ripe fruit. What this suggests is that, despite its frequency, hunting is not an essential economic activity for chimpanzees. Indeed, a recent appraisal of chimpanzee hunting in a forest in Uganda found that hunting was consistent seasonally, and was thus not done to make up dietary shortfalls when preferred kinds of food were scarce. Meat may have made a nice dietary supplement, but it did not seem to be critical to the nutritional needs of the chimpanzees. So why go to all the trouble? One possibility was that the sharing of meat by successful male hunters gave them preferential access to females, and therefore a reproductive advantage. Here the evidence is equivocal, and observations from different localities have varied. Generally, sexually receptive females are successful in obtaining a share of hunted meat only about a third of the time, doing no better than anyone else. And when they do get a share, it is as likely to be after as before copulation. On the other hand, one study in West Africa showed that, over the longer term, males got more sex from females with whom they had shared meat. Still, a consensus does seem to be developing among chimpanzee researchers that chimpanzees hunt colobus at least principally in pursuit of a larger social goal: to obtain meat to be shared, mostly with other males, in ongoing alliance-building exercises. This makes considerable sense, for chimpanzee societies are fluidly hierarchical, each male's position in society (and, of course, his reproductive advantage) at a given moment being determined not solely by his strength or his temperament, but by the state of his coalitions with others.

SPECIALIZED BIPEDAL APE BEHAVIORS

Hunting is one thing, and tool-making is another. Animal carcasses were butchered early on in hominid history, but this says nothing about how the carcasses themselves were obtained. Certainly, the very earliest stone tools don't look at all like the kinds of implements that would have been used in killing large animals. If the early hominids did have a substantial meat component in their diet, it seems overwhelmingly likely that many of the animals providing the meat were small enough to be captured by chasing and cornering. Those hyraxes, for example, or small vertebrates like lizards. The only plausible alternative to this (apart from that sneaky leopard-kill stealing) is that significant protein came from the larger prey of full-time predators that were temporarily driven away from their kills through some form of hominid aggression. Given that the early hominids were small bodied, slow moving, and not fearsome in the least, and that non-fresh carcasses may be deadly, there is only one obvious possibility in the latter event: that hominids had already learned to accurately throw heavy objects.

Throwing seems like a natural thing to us today, and indeed such sports as baseball depend on it; but in reality it's another of our unusual qualities. Today we are the only precision thrower out there. A camel may be able to spit in your eye, but for all the strength in their arms, chimpanzees can't throw a rock very far even if their accuracy up close seems reasonably good. They may have a fearsome feces-flinging reputation among zoo-goers; but they don't have the projectile abilities you'd want to depend on in a life-and-death situation. And this of course is just what you'd be in, if you were trying to chase carnivores off a kill by hurling rocks at them in the process known to paleoanthropologists as "power scavenging." Throwing accurately involves exquisite harmonization between the hand and the eye, and the ability to string together

In this imaginative reconstruction we see a group of early tool-using hominids out on the African savanna in the period before two million years ago. We do not know exactly how hominids like these obtained the mammal carcasses they butchered, but the figures in the left middle-ground hurling rocks at a pack of hunting dogs suggest the very hazardous process of power scavenging: driving the primary predators temporarily from their kills, while limbs and organs are removed for consumption at safer spots. Such refuge might have been provided by the forest in the background. Meanwhile, in the right middle-ground two of the hominids are seen using a stick to dig up tubers, also a resource for the earliest hominids that ventured out into the expanding grasslands. © '95 J. H. Matternes.

a whole sequence of actions based on an instinctive assessment of what is needed. This is no small feat of neuromuscular coordination; and we have no *direct* evidence that any hominids could manage it before they began to make stone tools of the kind that were evidently hafted as the tips of missiles.

Still, a good degree of hand-eye coordination is an essential component of stone tool making; and this suggests that despite their archaic body proportions the early stone tool makers might have been able to cultivate throwing skills good enough to help them obtain meat at least occasionally. Carnivore kills, though, are hardly the kind of food source any hominid would want to depend on. You'd certainly have to be very highly motivated to make a living this way; and if chimpanzees are anything to go by, it's not at all a sure bet that our ancestors went after meat initially because they were starving—though once they had started scavenging fresh kills with any regularity, it's fairly easy to see how they might have become dependent on this practice. But if they did take this hazardous dietary route—and we need to bear in mind that from the beginning they were butchering large animals—it would be a powerful indication that they lived in large aggregations, since a small group of tiny hominids tossing rocks at lions, or sabertooths, or giant hyenas, would have got pretty short shrift. And there are other reasons, which we'll look at shortly, to suggest that early hominids lived in large groups.

Meanwhile, those butchered carcasses also suggest much about both the early tool-wielding hominids themselves and the cognitive level they had achieved. Rock suitable for making tools does not occur everywhere on the Rift Valley landscape over which the early hominids roamed. And when they needed to butcher a large animal carcass, the evidence is clear that they assured the availability of good tool-making materials by carrying appropriate rocks around with them. Especially after we pass the two-million-year point, it becomes reasonably common to find the fossilized remains of cut-marked butchered carcasses with ancient stone tools scattered around and even within them. Characteristically, the tools themselves were not made from stone naturally available in the immediate area; sometimes the nearest natural source was often several miles away. In such cases the fine-grained rocks needed for tool-making must have been carried in from at least that far afield. What's more, they were not brought in as slimmed-down finished tools. We know this because the tool-making process didn't just produce one sharp tool after another. A single cobble might have produced two or even several cutting flakes, but in the process a lot of "debitage"—unusable fragments of stone—also resulted. And archaeologists have repeatedly pieced together entire cobbles from both useful and useless fragments found at a single butchery site. Not only has this laborious process of reconstruction told them a lot about how the stone tools were made, but it also provides clear evidence that the heavy cobble had originally been carried in complete—evidently in anticipation of being needed for tool production. And it hardly seems likely that hominids would have lugged weighty chunks of rocks around for miles, if they were rarely going to use them.

This kind of anticipation and foresight is different from anything we see in chimpanzees. Certainly, they hunt. But they usually do it on a situational basis, according to opportunities that present themselves. And if they need an implement to perform an activity with, they fashion it from available materials on the spot. The early stone tool makers, on the other hand, seemingly set out knowing precisely what they were going to do—whether hunting, or power scavenging, or whatever—and what they would need to do it with. They also understood the properties of materials, and how to work them, in a way that chimpanzees don't. This is already enough to tell us that the early hominids had taken a cognitive leap of some kind compared not only to their Pliocene relatives, but to modern apes as well. Clearly, in their time the tool-wielding australopiths, and plausibly also their immediate non-technological ancestors, were by a substantial margin the smartest creatures around.

Sadly, at present there is much less to say about these ancient precursors than we'd like, although it's a good bet that they were very cooperative creatures. But if you crammed four hundred chimpanzees into the back of an airplane and flew them from New York to Tokyo, there's little doubt that on arrival you'd find that the chimps had all massacred each other. Chimpanzees are highly social beings by any standard, but they do not have the special kind of sociality that it takes to live in a world as packed with people as ours is today. We certainly didn't acquire this particular form of sociality in response to modern crowded conditions, for our population explosion is recent; indeed, hominids have typically been very thin on the ground, at least over the past two million years or so. Perhaps, then, we should seek earlier in our evolutionary history for the biological underpinnings of our peculiar social propensities. And one suggestion is that we should look to the biological role and environmental preferences of the early bipeds.

EARLY SOCIETIES

So far we have discussed much about chimpanzees and hunting, and this is certainly a reasonable thing to do if we want to place our ancient ancestors in the context of our (and their) closest living relatives—creatures that we can actually watch going about their business in the natural world. It is also true that an emphasis on hunting is deeply embedded in paleoanthropological tradition. Indeed, in the 1950s Raymond Dart, who described the very first australopith fossil (actually, the infant victim of a predatory eagle) back in 1925, was dramatically proclaiming that "the blood-bespattered, slaughter-gutted archives of human history" were a direct reflection of "this common bloodlust differentiator, this predaceous habit" of mankind's earliest ancestors.

However, while we are undoubtedly the world's top predator today, my colleagues Donna Hart and Bob Sussman have recently emphasized just how misplaced this focus on hunting in our early evolution is. They point out that we are not simply super-chimpanzees; and that for all their evolutionary nearness to us, chimps retain all the instincts of forest animals, even where they spend a lot of their time in thinly wooded settings, as at Fongoli. According to Hart and Sussman, what most fundamentally differentiated our very ancient ancestors from chimpanzees is that, unlike the living apes, they adjusted their entire way of life to the exploitation of forest-edge and woodland settings. We see this in their bipedality, in their teeth, in their geochemistry, and in a host of other features. The ecological move to these more open environments brought with it new opportunities for hominids, as well as extraordinary future possibilities; but it also came at a huge immediate cost. This penalty was, of course, vulnerability to woodland predators. It is impossible to overstate the significance of this new factor: no new force could have had anything close to the impact on small-bodied bipeds, venturing away from their ancestral habitat, that the ubiquity of predators must have had.

Given this inescapable reality, Hart and Sussman suggest that we should probably not look first to our very closest extant relatives for clues as to how our earliest relatives lived. Instead, we would be better off seeking indications from environmentally similar primates such as macaques and baboons. Even if more distantly related to the australopiths than chimpanzees are, these primates have made a similar ecological commitment to living with both the advantages and disadvantages of the expanding new habitat mosaic. True, it's probably about 25 million years since we shared an ancestor with them, but our basic primate biology is similar, as is our ecological bias. Moreover, the fossil record shows that our forebears in the period before about 2.5 million years ago were not much bigger than largish baboons. One big difference, though, was in the size of the canine teeth. Male baboons, in particular, have fearsome, slashing upper canine teeth with razor-sharp back edges, a defensive feature that was conspicuously lacking in our own precursors. And the quadrupedal baboons are far fleeter of foot (indeed, their ground-favoring relatives the patas monkeys can hit almost 40 mph when they have to). The australopiths were thus significantly more vulnerable in open habitats than the terrestrial monkeys are, and the predatory pressures on them would have been concomitantly greater.

As you'd expect from animals that are at least partially committed to the savanna, baboons and macaques are omnivorous, exploiting the resources of the grasslands as well as of the forest, although they are also modestly tied to sources of water. But although they move well out into the grasslands to forage during the day, they commonly cluster for protection at night in trees or on cliff faces. And like other conspicuous species vulnerable to predation, they live in very large groups consisting of multiple males and females of all ages. After all, the more eyes and ears there are, the more likely it is that someone will spot a faraway predator and raise the alarm. No surprise, then, that these monkeys are also quite vocal. Often the groups forage and move around in such a way as to keep the reproducing females and young in the physical center, while the more expendable young males remain at the vulnerable periphery, where they can also function as sentinels. Since the groups are large, they are well structured and organized, with complex individual relationships among the members. This orderliness is unlike what we see in chimpanzees, which live in groups that lack rigid spatial structuring—even though within them inter-individual relationships are yet more complex.

There is plenty of evidence, mostly in the form of fractured bones and carnivore tooth-marks, that early hominids were frequently preyed upon; and the indirect evidence of habitat and body size and anatomy speaks to the same thing. Hart and Sussman thus reasonably conclude that the very early hominids would have had the social characteristics not of hunters, but of *prey* species. Our ancestors were the hunt*ed*, not the hunt*ers*; and these authors believe that much in our modern behavior still reflects this. We'll return later to the subject of our behavioral heritage; but meanwhile, Hart and Sussman identify seven strategies used by terrestrial monkeys that they believe early hominids would almost certainly have employed in their vulnerable new niche:

1. Live in large groups, from 25 to 75 individuals. There *is* safety in numbers. Perhaps influenced unconsciously by the knowledge that the human nuclear family is usually small in our own society, and more consciously by the demographics of predators, paleoanthropologists have tended to assume that early human groups were limited in size. As we've seen, bipedality has been linked with pair-bonding, and the large size differences between males and females of *Australopithecus afarensis* have invited comparison with gorillas, which usually live in groups of under 20 individuals that are dominated by a "silverback" male. For vulnerable prey species, though, significantly larger social groups would plausibly have been the norm.

2. Be versatile. Don't put all your eggs in one basket, as it were, but use all the environments and substrates available to you. We know that this rule applied to the early hominids, which combined bipedality on the ground with significant agility in the trees. Monkeys mostly achieve this versatility by staying small and generalized; early hominids did the same thing by combining seemingly contradictory specializations. It seems pretty clear that the hominid "have your cake and eat it too" locomotor strategy was not a transitional adaptation by creatures who were caught in the act of descending from the trees to the ground. They thrived on this way of life for many millions of years; and even

conditions that we perceive in retrospect as "intermediate" must have been entirely functional in their day. Their body form indicates that early hominids were expressing a stable strategy, the diverse components of which, including the terrestrial leg and pelvis and the arboreal shoulder girdle and arms, seem to have been well accommodated to environmental necessity—despite the vulnerabilities inherent in the new way of getting around on the ground.

3. Be flexible in your social organization. Avoiding predators is fine, but it shouldn't come at the cost of starving yourself. On the savanna especially, the kinds of resources that a primate may readily access tend to be scattered, and are rarely abundant in one place. The large social unit should thus break up into smaller ones to allow more effective foraging for scarce resources, but all must be ready to re-coalesce into the larger group when real danger threatens.

4 and 5. Although males as a category are more reproductively dispensable than females are, have more than a single male in the social group at all times, even when smaller subgroups are roaming around. And use those males as sentinels, especially where males are larger than females and better able to discourage predation. Upright locomotion actually may help here, because it makes individuals appear larger to predators, and may fail to trigger an attack response in the way that horizontal silhouettes do.

6. Select your sleeping sites carefully. Assemble the group at night in trees or other places of comparative safety, and during the days stay as much as possible in areas of comparatively dense vegetation. When moving through open areas, maintain the largest possible group size.

7. Be smart. The better you are able to read and interpret the environment, the safer you will be. The better you communicate, the more effectively all members of the group will be able to avoid predators. Significant increases in hominid brain size—and, presumably, in intelligence—may not have really progressed until our precursors had been outside the dense forests for millions of years, and our own genus *Homo* had emerged. But the change of environment initiated by the first bipeds may well have been a critical enabling factor, setting the scene for later developments.

These seven strategies certainly do not add up to a full portrait of our ancient ancestors as socioeconomic beings. At present we can be confident only that our remote forebears adopted two of the strategies just listed: versatility, and use of the trees for shelter. The rest are just a best guess, based on what related forms do in similar circumstances. But even if this listing falls short of a characterization, there is a compelling case to be made that even today, humbling as it may be, the World's Top Predator bears the scars of its lowly beginnings as favored items on the menu of a whole array of carnivores.

THE INTERIOR WORLD

We are beginning to form a picture of the australopiths, hazy and incomplete though it might be. They were small-bodied upright bipeds, with considerable tree-climbing abilities: creatures that moved between the forest and more open environments where they would have lived in large social units for protection. They had complex social lives, based on intensive cooperation and a special form of sociality expressed by groups embracing many individuals of each sex and age. They were highly vocal; and, by analogy with living apes (presumably the best parallel in this respect), they would have had a vocabulary of several dozen distinct utterances, each expressing one of a range of different situations or emotional states. We can confidently say that these remote forebears were generalist omnivores, exploiting what both the forest and the savanna had to offer; and in this way they differentiated themselves even from modern savanna apes, who seek forest-type resources wherever they may be. The archaic hominids lived at least part time in a dangerous and challenging environment. And though they had brains not much bigger than those of apes of comparable body size, at some point they began to make stone tools and carry around the materials necessary for their manufacture, indicating a level of cognitive complexity beyond what any ape has yet demonstrated. The tools and the carcasses they dismembered provide the first evidence we have for the consumption by hominids of animal fats and proteins, although by analogy with chimpanzees it seems reasonable to conclude

that flesh-eating and meat-sharing may have been an established behavior long before.

What made the intellectual leap to stone tool making possible is not
something we can hazard with any confidence at this point, although
the refinement of motor skills and of higher cognitive functions almost
certainly went hand in hand. But the fact that the first stone tools—the
first step in an epic transformation—were made by creatures whom we
can—with reservations—characterize as "bipedal apes," inaugurates
a pattern that we will see recurring repeatedly over the entire span of
hominid evolution: new technologies (reflecting new and more complex
behaviors) do *not* tend to be associated with the appearance of new
kinds of hominid. It was old kinds of hominid that started to do new
things, even though those new things always seem to indicate a step up
in cognitive complexity.

We will return to the typical hominid pattern of innovation. But
first, it might be interesting to ask if we are in a position to form any
impression at all of what kind of sense of the world around them—or of
themselves—those bold small-bodied bipeds possessed. We can infer a
lot about how their lives might have looked to an observer. But did they
share with us any aspects of the unique modern human form of inner
experience? There is no way to answer this question with any precision;
but one thing that we can do is to set an approximate baseline by looking at other organisms and asking what we demonstrably share with
them, and by extension with the early hominids.

One obvious issue to start with is the sense of self. In the very broadest of meanings, every organism has a sense of itself versus the other.
From the simplest unicellular creature on, all living things have mechanisms that allow them to detect and react to entities and events that lie
beyond their own boundaries. As a result, every animal may be said
to be self-aware at some level, however rudimentary its responsiveness
to stimuli from outside might appear. On the other hand, human self-
awareness is a highly particular possession of our own species. We human beings experience ourselves in a very specific kind of way—a way
that is, as far as we know, unique in the living world. We are each,
as it were, able to conceptualize and characterize ourselves as objects
distinct from the rest of Nature—and from the rest of our species. We

consciously *know* that we—and others of our kind—have interior lives. The intellectual resource that allows us to possess such knowledge is our symbolic cognitive style. This is a shorthand term for our ability to mentally dissect the world around us into a huge vocabulary of intangible symbols. These we can then recombine in our minds, according to rules that allow an unlimited number of visions to be formulated from a finite set of elements. Using this vocabulary and these rules we are able to generate alternative versions or explanations of the world—and of ourselves. It is this unique symbolic ability that underwrites the internalized self-representation expressed in the peculiarly human sense of self.

In between the two ends of the spectrum, linking the primordial and the symbolic styles of self-awareness, there presumably exists a near-infinite array of states of self-knowledge. Yet because alien cognitive states are among the few things human beings find it impossible to imagine, let alone to experience, any discussion of such intermediate forms of self-knowledge—such as that possessed by our early ancestors—is fraught with huge risks of anthropomorphizing. When we try to understand how other organisms comprehend particular situations, or their place in society, or indeed their place in the world, our tendency is always to impose our own constructs. The temptation is to assume that beings of other kinds are seeing and understanding the world somehow as we do, just not as well or as fully. Yet the truth is that we simply cannot know, still less *feel*, what it is subjectively like to be any organism other than ourselves, modern *Homo sapiens*.

The extraordinary human cognitive style is the product of a long biological history. From a non-symbolic, non-linguistic ancestor (itself the outcome of an enormously extended and eventful evolutionary process), there emerged our own unprecedented symbolic and linguistic species, an entity possessing a fully-fledged and entirely individuated consciousness of itself. This emergence was a singular event, one that involved bridging a profound cognitive discontinuity. For there is a *qualitative* difference here; and, based on any reasonable prediction from what preceded us, the only reason for believing that this gulf *could* ever have been bridged, is that it *was*. And since that extraordinary event self-evidently did take place, the question becomes one of where and how. To answer this, though, we need to establish that starting point. This is no easy task, and how difficult it is in practice is well illustrated by the investigation of self-recognition.

Back in the mid-nineteenth century, Charles Darwin placed a mirror on the floor between two orangutans housed at the London Zoo. He recorded a variety of reactions made by the orangutans to their reflections, but was vague as to what, if anything, he had specifically concluded from the experiment. There the matter rested for almost a hundred years, until the cognitive psychologist Gordon Gallup, noting that the norm among animals was to treat mirror images as other individuals, carried out a more controlled test. Gallup exposed two juvenile chimpanzees to full-length mirrors for several days, and watched how they responded to the images of themselves they saw reflected. Over this period, self-directed behaviors increased, while social reactions to the mirror images declined, suggesting that the individuals were learning to recognize the images as themselves. The chimpanzees were then anesthetized, and red marks were applied to their faces. Once they were reintroduced to the mirrors, self-directed behaviors intensified, many of them aimed at the marks. In contrast, marked chimpanzees without prior mirror experience failed to respond in this way, suggesting that self-recognition had indeed been learned by the first group during the habituation period. Similar testing of macaques produced contrary results, implying to Gallup that these monkeys lacked the chimpanzees' capacity for learning self-recognition.

Since Gallup's pioneering study, the "mirror test" has become the standard yardstick for self-recognition among vertebrates, and a wide variety of species has been tested. Human beings naturally need to learn mirror self-recognition (MSR) just as the chimpanzees did; but adults to whom sight has been restored do so quickly, and most human infants can manage the trick at 18 to 20 months of age. Young apes develop more rapidly than human children in many respects, but studies building on Gallup's original have shown that MSR is rare among chimpanzees under eight years old: it is basically an adult ability. By now, MSR has been demonstrated not only in chimpanzees but also in bonobos, orangutans, and gorillas, although not all tested individuals of these species have shown it. Outside the human–great ape group MSR is evidently extremely rare among vertebrates (though elephants, dolphins, and certain birds may display it); and to the extent it occurs, different underlying mechanisms are almost certainly at work than those operating in great apes and humans. But although its expression in apes and humans is almost certainly a unique property of this group, uncertainty still remains as to what exactly MSR is revealing: what it means in terms of the precise aspects of consciousness that the approach explores.

An alternative avenue to understanding the sense of self in nonhuman primates was thus taken by the monkey researchers Robin Seyfarth and Dorothy Cheney, who adopted the psychologist William James' distinction between the two components of self-awareness: the "spiritual" (one's "psychic faculties and dispositions"), and the "social" (knowledge of being one of many distinct individuals embedded in a group). Like human beings, monkeys are intensely social, and Seyfarth and Cheney looked at how individual vervet monkeys and baboons appeared to understand their places in the social hierarchy. The reasonable assumption here was that a primate cannot exhibit a sense of "them" without also possessing a sense of "I"; and, from looking both at kin relations and at the dominance hierarchies to which the monkeys belonged, Seyfarth and Cheney concluded that they did indeed recognize other group members as individuals, behaved toward them in appropriate ways, and hence appreciated their own individuality vis-à-vis their fellows. This seemed to indicate that on some level they had a sense of the social self.

On the other hand, this kind of self-awareness was clearly different from that of human beings. For, while they are certainly able to behave appropriately in complex social settings, vervets and baboons are, as far as one can tell, unaware of the knowledge that allows them to do so. In Seyfarth and Cheney's words, they "do not know what they know, cannot reflect on what they know, and cannot become the object of their own attention."

No observer would deny that great apes possess more complex cognitive and behavioral repertoires than monkeys do. Still, it is far from clear just how far they exceed them in these last respects, and particularly in the ability for self-reflection. Some great apes, like our friend Kanzi, are highly adept users of symbols in experimental situations. They can recognize and respond precisely to words and even to combinations of words, and they can choose visual symbols adroitly on a computer screen. But whether this means that they are also able to manipulate such symbols mentally in such a way as to produce objective images of themselves is doubtful. In general, the apes' use of symbols seems to be additive: they can comprehend short strings of concepts ("take," "red," "ball," "outside"), but they do not recombine them according to mental rules to produce new notions: ideas of the possible, rather than of the observed. The chimpanzee manner of dealing with symbols is thus is inherently limited, since lengthening lists of symbols rapidly become confusing, and ultimately meaningless.

Daniel Povinelli, a distinguished researcher of chimpanzee cognition, proposed a few years ago that a fundamental distinction between the ways in which chimpanzees and humans view the world is that, while humans form abstract views about other individuals and their motivations, "chimpanzees rely strictly upon observable features of others to forge their social concepts. . . . [They] . . . do not realize that there is more to others than their movements, facial expressions, and habits of behavior. They [do] not understand that other beings are repositories of private, internal experience." It also implies that individual chimpanzees do not have such awareness of themselves, either. They *experience* the emotions and intuitions that arise in their own minds; and they may act on them, or suppress them, as the social situation demands or permits. But, just as in Povinelli's words they "do not reason about what others think, believe and feel . . . because they do not form such concepts in the first place," it seems legitimate to conclude that this exclusion also applies to self-reflection. Because, if individual chimpanzees lack the ability to perceive that others have internal lives, it is highly probable that they also lack equivalent insight into their own interior existences.

Profound as it is, this cognitive difference between us and them may not always produce radically distinctive observable behaviors; and indeed the ways in which chimpanzees and humans behave sometimes appear strikingly similar. Still, we should be wary of overstating these similarities. The behavioral resemblances we perceive are conditioned by an enormously long shared evolutionary history, and by the resulting structural similarities. But, as Povinelli would point out, similar observable behaviors may also hide mental processes that differ greatly in form and complexity.

So, for all the manifold talents that chimpanzees possess, that cognitive gulf still yawns. Among all those organisms that we can study in the world today, it appears that only modern human beings show "spiritual self-awareness" in William James' sense; and even his "social self-awareness" appears to differ dramatically in quality between humans and nonhuman primates. Still, even though a lot of evolutionary water has flowed under the bridge on both sides since human beings shared an ancestor with any ape, most authorities find it reasonable to conclude that cognition of the kind we see among chimpanzees (and among other great apes as well) provides us with a reasonable approximation of the cognitive state from which our ancestors started some seven million years ago. To return to Povinelli's words, one may reasonably assume that those ancestors were "intelligent, thinking creatures who deftly attend[ed] to and learn[ed] about the regularities that unfold[ed] in the world around them. But . . . they [did] not reason about unobservable things: they [had] no ideas about the 'mind,' no notion of 'causation.'" In the human sense, they had as yet no idea of self. This is a very plausible characterization of our lineage's cognitive starting point; but at the same time it more or less exhausts what can usefully be said on this subject, based on our existing knowledge of comparative cognition.

The next question is, of course, to which of our known ancestors does Povinelli's characterization apply? In all probability, if we were able to directly observe the very early hominids we met in chapter 1 we would find that Povinelli's depiction fit them well enough; and we have no compelling reason for believing that it would not also have applied very broadly to the earliest *Australopithecus afarensis*. Still, if it was indeed *A. afarensis* (or something very like it) that introduced stone tool making, as the Dikika (and Bouri) evidence seems to suggest, then we have to reconcile the Povinelli viewpoint with the significantly advanced cognitive achievements of those first stone tool makers. For there is no doubt that the first hominids who made stone tools and used them for butchering carcasses were displaying evidence of an entirely new and radical way of interacting with the world around them; and there is no reason to believe that this innovation might not have had internalized effects too. The simplest and most plausible way of explaining this apparent discrepancy is to suggest that the cognitive *potential* to make stone tools was born in the large genetic alteration that must have been involved in the acquisition of the new and radically different bipedal body form; and that this potential lay dormant for some time before being expressed in the invention of stone tool making.

This scenario is not as far-fetched as it first seems if we remember that, as I've already suggested, innovations are not acquired directly as *ad*aptations, but must instead arise as *ex*aptations that must be co-opted *post hoc*. Examples of dramatic physical exaptations making possible huge later behavioral leaps are no rarity in evolution: birds, for example, possessed feathers for years before recruiting them for flight. And the ancestors of the tetrapods, the four-footed land-dwellers, initially acquired their limbs in an entirely marine setting.

We will return to this subject, because the one we've just discussed is far from the last spectacular example of exaptation as an indispensable agent of innovation in hominid evolution. But meanwhile, knowing that doesn't help us much in understanding potential changes in early hominids' perception of themselves and the world around them. Because although the toolmakers were displaying an insight into the potential to modify that world in ways that would eventually have literally earth-shattering consequences, we have no idea whatsoever how this new ability affected, or was reflected in, other areas of their behavior and experience. All we can say is that they were behaving in entirely unprecedented ways: ways that underwrote all the manifold changes that were to come, but into which we cannot yet read evidence of any of the other attributes that make us feel so different from all other creatures.

FOUR

AUSTRALOPITH VARIETY

I t is reasonable for many reasons to start our discussion of the australopiths with *Australopithecus afarensis*. For a start, this is widely believed to be the "stem" species from which later australopiths were derived. And so far it is by a clear margin the best-known of all the early hominid species, so it acts as the perfect foil for discussing virtually all the issues of early hominid lifestyle that the australopiths raise. But it should never be forgotten that *A. afarensis* was only one species among many in the early hominid spectrum over the period between about 3.8 and 1.4 million years ago. This places hominids very comfortably within the pattern of diversification that we see in all similarly successful mammal groups. One of the best-documented evolutionary phenomena is "adaptive radiation"—the rapid multiplication of species that occurs when an organism enters a new "adaptive zone" in the way the early bipeds did, and the descendants of a single hardy adventurer diversify to exploit all of the new possibilities available to them. This has happened over and over again, and there are few better examples of the phenomenon than what occurred after the first hominids committed themselves to the ground. Chimpanzees may do quite well in woodland settings; but they do so by employing their existing attributes in slightly new ways, rather than by making the kind of radical change that the

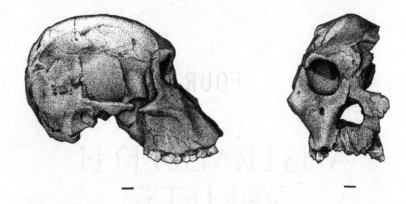

The cranium of a gracile australopith, Australopithecus africanus, *from the South African site of Sterkfontein. The specimen is known as Sts 71, and it is probably around 2.6 million years old. Drawing by Don McGranaghan.*

early hominids made. This in turn helps to explain why the range expansion of chimpanzees into savanna environments has had no effect on their diversity—the savanna chimpanzees remain members of the same species as their forest-dwelling relatives. Early hominids, on the other hand, made a physical rather than a simply behavioral response to life (at least part time) on the ground; and this opened up a host of new opportunities to them which they exploited to the full.

The very first australopith fossil ever discovered was found in a lime mine on the high veld of South Africa in 1924, and in the late 1930s others began turning up not too far away at similar sites. Quickly it was realized that at least two radically different kinds of small-brained early hominid were represented among these fossils: a lightly built form known as *Australopithecus africanus* ("southern ape of Africa"), and a relative with a more massively constructed skull called *Paranthropus robustus* ("robust near-man"). Both had small cranial vaults containing brains that, at best, were only slightly larger than those of modern apes in comparison to their assumed body size. However, while *A. africanus* had boasted a set of teeth that were proportioned very much like those of *A. afarensis,* the teeth of *Paranthropus* were very different. The incisors at the front were small, and the canines behind them were likewise reduced. But the premolars and molars were broad and flat, together forming an impressively specialized grinding machine. The teeth were set in a face that was relatively short from front to back (because those front teeth didn't take up very much space) but that was massively built both to accommodate the huge molars and to absorb the stresses they generated. So large were the muscles that moved the jaw that a tall ridge (the "sagittal crest") ran backward along the center of the braincase to allow additional attachment room for large temporal muscles—just as we see among male gorillas today.

For want of postcranial bones, nobody knew quite how big-bodied *P. robustus* individuals would have been, but quickly a dichotomy was established between the "robust" *Paranthropus* types and the "gracile," or slender, *A. africanus*. Neither did anybody know exactly how old these early hominids were, although from the accompanying faunas it was guessed that the graciles were generally older than the robusts. Now we have a pretty good idea that the South African graciles (recently augmented by the finding at a site called Malapa of a fabulous trove of fossils representing a new and in many ways advanced species, *A. sediba*) bracket the period between about 3.0 and 2.0 million years ago or so, and that their robust counterparts date from around 2.0 to 1.7 million years ago.

At one gracile site, Sterkfontein, a slightly older find deserves special mention, if only because of the extraordinary circumstances of its discovery. The majority of hominid fossils from Sterkfontein date from about 2.5 million years ago or so, but in an underground section of the site some cave deposits are exposed that may be over three million years old. Rummaging around in a collection of bones dynamited decades ago from these older cavern deposits, the paleoanthropologist Ron Clarke came across parts of an ankle joint and foot that his practiced eye immediately recognized as hominid. Noticing that the break in the shin fragment that provided the upper part of the ankle joint looked fresh, he asked his colleagues Stephen Motsumi and Nkwane Molefe to go into the vast gloomy cavern and look for the counterpart whitish cross-section—smaller than a dollar coin—embedded in the gray cave wall, truly a search for a needle in a haystack. But miraculously, those eagle-eyed searchers promptly located the ring of bone Clarke had predicted would be there. They then began the long labor of extracting the rest of the skeleton (dubbed "Little Foot") from the rock-hard matrix in which it had been entombed for the last three million years—a process that isn't yet complete as of the time of publication. Enough of the skeleton has been exposed, however, to reveal that it isn't quite like the *A. africanus* fossils from the later layers at Sterkfontein, and that it may represent a species ancestral to a second and as-yet-unnamed species that also comes from those later deposits. At the single site of Sterkfontein, therefore, we are seeing evidence of unexpected diversity among the gracile australopiths alone.

Just looking at the teeth of the robust and gracile australopiths, you would immediately imagine that they were eating totally different things. The teeth of *Australopithecus africanus* look much more generalized, like those of an opportunistic fruit-eater that also availed itself of everything else in sight—just as you would expect from a relative of the ancestral bipeds. The teeth of the robusts, on the other hand, look like the specialized grinding apparatus of a dedicated feeder on hard and maybe gritty vegetal materials, such as the roots, tubers, and other items typical of open environments. But studies of the way those teeth actually wore as a result of chewing showed that things may not have been quite so simple. Examination of the dental wear surfaces of both graciles and robusts, under very high magnification, showed that there had in fact been a great deal of overlap in what all the hominids had consumed overall, and that any significant dietary differences were probably confined to times of the year when environmental productivity was low. At such times, the different kinds of hominid may have resorted to different "fallback" foods: hard and brittle in the case of the robusts, tough but more yielding in the case of the graciles.

The notion that robusts and graciles in South Africa had largely similar diets also emerges from studies of stable carbon isotopes. These showed a lot of variability among samples, but yielded basically identical patterns for both kinds of australopiths—with a strong C_4 signal in each case. Researchers believe that the C_4 indications result largely from the consumption of animals such as hyraxes, cane rats, and young C_4-plant-consuming antelopes—though they do not rule out the possibility that some of it was due to consumption of grasses, most probably in the form of their rhizomes (underground runners). Interestingly, although it's known that South African climates and environments fluctuated quite considerably during the tenure of the australopiths, the differing carbon-isotope ratios observed do not correlate with time. Both kinds of australopiths thus seem to have maintained their generalist dietary proclivities even as the habitat changed around them.

All this strongly supports the idea that the success of all or most of the early hominids was due to an opportunistic strategy: they all ate everything available, rather than restricting themselves to particular types of resource. Only under stress might they have resorted to significantly different foods. It is this ubiquitous increase in dietary breadth that seems to have distinguished the australopiths most strongly from chimpanzees, which as we've seen maintain a strong preference for forest-type products even when they range into the savanna. The generalist proclivity that we see among australopiths also has implications for the evolutionary origins of the diversity we see among them: perhaps we should look at their sometimes huge anatomical differences as being the result of the fixation of chance novelties, rather than as the result of fine-tuning adaptation over the eons.

Crude stone tools are known from deposits at Sterkfontein dating close to two million years ago; and similar implements up to about 1.8 million years old are also known from a nearby robust australopith site called Swartkrans. This latter has also yielded some polished pieces of bone that bear scratches identical to those you would make today if you tried to dig up roots and tubers using similar improvised tools. Both sites have additionally produced rare fossils that have been attributed to our own genus *Homo*, and that were assumed to have been the remains of the tool-makers. Still, the new evidence from Ethiopia makes it much easier to entertain the more obvious explanation that both gracile and robust australopiths were making and using stone and other tools at least as far back as two million years ago. This fits nicely with the interpretation of hand bones from Swartkrans that are almost certainly those of *Paranthropus* and that show anatomical features compatible with high manipulative abilities. In known postcranial features the South African graciles (except Malapa) look very much like Lucy (we have very little evidence for the robusts). All in all, the emerging picture from South

Africa is looking similar to the one we are deriving from the later part of
the tenure of *Australopithecus afarensis* in Ethiopia.

EAST AFRICA

South Africa was the first region of the world to provide evidence of very
early hominids. But from the early 1960s on, eastern Africa grabbed
the limelight. In 1959 the legendary Louis and Mary Leakey announced
the discovery at Tanzania's Olduvai Gorge of the fossil cranium of a
"hyper-robust" australopith that they termed *Zinjanthropus* (for the old
Arab-ruled "Zinj" empire along the East African coast), or more fondly
"Nutcracker Man," on account of the flat and hugely expanded chewing teeth that totally dwarfed its tiny incisors and canines. Nowadays
this specimen is assigned to the species *Paranthropus boisei* (named for
a benefactor of the Leakeys' research), and is dated to 1.8 million years
ago. The Leakeys had been finding primitive stone tools at Olduvai for
years and, as a devotee of the "Man the Toolmaker" notion, Louis was
already convinced that these crude implements must have been the work
of an early member of the genus *Homo*. Having spent some three decades periodically prowling the Gorge under the blistering African sun in

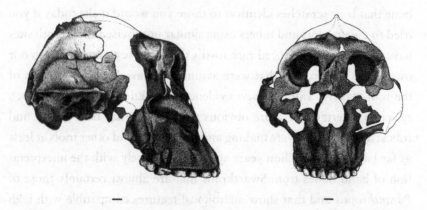

*Cranium of Olduvai Hominid 5, a.k.a. "Zinjanthropus," a robust australopith
of the species* Paranthropus boisei, *from Olduvai Gorge, Tanzania. 1.8 million
years old. Drawing by Don McGranaghan.*

search of the toolmaker, the Leakeys were elated to find any hominid at all; but they were naturally a bit disappointed when the one that eventually turned up was not something they had any hope of squeezing into the genus *Homo*.

Still, they did not have to wait long to find what in their terms was a much better candidate for the honor of being the long-sought Olduvai Toolmaker. For in 1961, at about the same geological level in the Gorge as the *Paranthropus*, Louis reported a lower jaw of a much more gracile hominid. Various people at the time noted striking similarities between the teeth of this specimen and those of *Australopithecus africanus*; but Leakey was undeterred in his search for early *Homo*, and a few years later he and some colleagues made the Olduvai jaw the holotype of *Homo habilis*, "handy man," named of course for its presumed manual skills. Thus began the paleoanthropological tradition of routinely assigning East African early gracile hominids not to the genus *Australopithecus*, but to our own genus *Homo*—a tradition that was only broken a decade and a half later, when the much earlier and more robust *Australopithecus afarensis* was first announced from Hadar and Laetoli. Those fifteen intervening years were very eventful ones in paleoanthropology.

In deference to tradition, we will return to the gracile Olduvai hominid and other such fossils in chapter 5, when we look at the evidence for the origin of the genus *Homo*. Meanwhile, though, the robust australopith from Olduvai became simply the first of many such fossils to be published from sites in Tanzania, Kenya, and Ethiopia. In the 1960s, expeditions to contiguous regions of southern Ethiopia and northern Kenya turned up evidence of fossil hominids in the time frame between about 2.6 and 1.5 million years ago. Many of these were "hyper-robusts." The earliest of them, in the 2.6- to 2.0-million-year bracket, came from the Omo Basin in southern Ethiopia, and were pretty fragmentary. Still, the jaw fragments were massive; and because they contained the same combination of huge, flat molars and tiny front teeth seen in Nutcracker Man, they were generally assigned to *Paranthropus boisei*—though one toothless jaw some 2.6 million years old received the name *Paranthropus aethiopicus*, for the country in which it was found.

Just to the south, on the eastern shores of Lake Turkana in northern Kenya, somewhat younger (1.9- to 1.5-million-year-old) robust australopiths began to show up in the late 1960s. These included one pretty complete, albeit toothless, robust skull that looks rather different from the Olduvai robust specimen, with a much broader and shorter face. Still, its dental proportions would have been basically similar and it, too, was assigned to the species *Paranthropus boisei*. Interestingly, we now know a frontal bone from East Turkana that looks just like its counterpart in the Olduvai cranium, and different from its Kenyan coeval. So the betting might be that we have more than one kind of robust australopith represented in the Turkana Basin at around 1.9 million years ago. Whatever the case, all of the East African hyper-robusts had similarly huge molar teeth that recent isotopic analyses have suggested were used to process large quantities of low-quality plant foods such as grasses and sedges. Apparently their diet was much more specialized than that of their South African relatives, and they may have been an exception to the rule of australopith omnivory.

One of the most exciting robust australopith finds at East Turkana was made in 1970, with the discovery of a partial skull of an individual who had been much smaller than the owner of the toothless robust skull, but who had belonged to the same species. Here at last was good evidence of sexual dimorphism—marked size differences between males and females—in robust australopiths. This find put an end, once and for

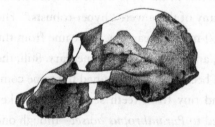

The "Black Skull," KNM-WT 17000, from Lomekwi, Kenya. Some 2.5 million years old, this is the most complete skull we have of the species Paranthropus aethiopicus, the earliest species of the "robust" australopith lineage.

all, to any lingering desire to categorize the robusts and graciles as males and females of the same kind of hominid, for it showed that female robusts didn't look like graciles, but rather like smaller versions of the robust males.

Work continued in East Turkana through the 1970s, but in the 1980s the focus of attention in the Turkana Basin moved to the western side of the lake, where slightly older fossil-bearing deposits were coming to light. In 1985 a famous specimen dubbed "the Black Skull" turned up. This specimen has many of the cranial characteristics of *Paranthropus boisei,* but its face is longer and concave in profile, and the braincase shows marked sagittal cresting at the very rear. The consensus rapidly developed that here was a form, ancestral both to *P. boisei* and to the South African robusts, which took the robust lineage back in time to 2.5 million years ago. Largely as a matter of convenience, it has been given the name of *Paranthropus aethiopicus* (after that jaw from Omo). At the other end of the time range, a 1.4-million-year-old skull from a place called Konso in southern Ethiopia has been identified as the last-known survivor of *Paranthropus boisei*—indeed, the last-known australopith of any kind. By that time, members of our genus *Homo* were all over the landscape; and indeed, advanced stone tools of the kind otherwise associated only with *Homo* fossils are also known from the Konso deposits.

Despite their strikingly different histories of interpretation, the eastern and southern African records are thus both beginning to give us a picture of the australopiths as a vigorous adaptive radiation, in which a whole variety of different early biped species were actively experimenting with ways to exploit their hominid heritage. Yet as far as we can tell, the basic pattern of adaptation—bodies light enough for agility in the trees, so fairly small, with a broad pelvis and short legs; omnivorous habits and with mobile forelimbs; relatively small-brained yet manually adept—persisted, even after the invention of stone tools announced that these revolutionary yet physically archaic (compared to us) creatures had developed an entirely new way of perceiving and dealing with the world. Theirs was truly a successful physical and behavioral strategy; and although it bridged the gap between the ancestral existence in the forest and the future occupation of open territory, it cannot be described as in any way "transitional" between the two. It was a way of life entirely unto itself, and it lingered on well after the point at which recognizable members of our genus *Homo* had come on the scene. Still, eventually the australopiths succumbed to those closer human relatives. Evidently, early *Homo* were invincible competitors, even if they were in competition for only a portion of the resources that the australopiths had exploited.

But even then the picture is not yet complete. In 2001, paleoanthropologists from the Kenya National Museums in Nairobi announced the discovery of a tantalizingly different fossil hominid, from sites to the west of Lake Turkana dating from 3.5 to 3.2 million years ago. The principal specimen is a crushed and cracked cranium that, even allowing for some distortion, is different from any of the other hominids known from the Turkana Basin. The chewing teeth seem to have been thick-enameled but quite small; and the skull itself is notably short-faced, hence the name *Kenyanthropus platyops* ("flat-faced Kenyan man") that the finders chose for their new fossil. Sadly, the condition of the specimen does not allow us to say much more about it; but the describers noticed distinct similarities to a skull known from much later deposits at East Turkana. Identified by the unromantic moniker of KNM-ER 1470 (its museum catalog number) this 1.9-million-year-old fossil (with a brain size a bit above the australopith range) had gained some notoriety in the early 1970s as the first skull to be discovered of *Homo habilis*—thus appearing to prove the reality of this species, since it didn't look like any known australopith. The unfortunate truth, however, is that this specimen also is so poorly preserved that it is difficult to know what to do with it. We'll look more closely at this fossil in the next chapter; suffice it to say here that placing it in *Kenyanthropus* seems for the time being to make eminent sense, if only because it sits very uncomfortably with either *Australopithecus* or *Homo*.

Clearly, a lot more was going on in human evolution at the end of the Pliocene than just the gradual refinement of a single central hominid lineage. The number of australopith species that we recognize from this late stage is multiplying, especially with the finding of forms such as *Australopithecus sediba* which appears to have a number of advanced features, especially in the pelvis, that does not at all have the dramatic sideways flare that we saw in Lucy's. This is also the time when the first claimed members of the genus *Homo* begin to appear, and whether or not they deserve this title it is becoming abundantly clear that this was a time of great evolutionary ferment among our precursors. The hominid stage was packed with actors, all pushing and shoving for the limelight; and the only thing we can be sure of is that the australopiths ultimately lost out.

FIVE

STRIDING OUT

People have referred to themselves as "human" since long before anyone had the faintest idea that our species is connected to the rest of the natural world by an extended series of long-vanished intermediate forms. And, at least until the notion came along that all living things are connected by ancestry, there was no compelling reason to define the term "human" precisely. This is why, a century before Charles Darwin published *On the Origin of Species,* the great Swedish natural historian Carolus Linnaeus was content to brush off *Homo sapiens* with the remark *nosce te ipsum* (know thyself). (Linnaeus gave us the system of classifying organisms that we use today, and one of his great innovations was to furnish diagnostic physical characters for every other species he named.) Clearly, Linnaeus and his contemporaries felt that our species is so distinct from all other living creatures as not to require formal delineation. And who could blame them for that? Given what was known of our zoological context in the eighteenth century, defining exactly what humans are simply wasn't a practical scientific problem, even though it had long entertained philosophers.

Today, though, it's different. For while we are the only "human" beings currently alive on Earth, we now know we have a whole range of close relatives—much closer than the apes—that are now extinct. What's more, those fossil relatives become more and more unlike ourselves as we trace them back in time. This naturally raises the matter of when

exactly our precursors became "human," a query that obviously leads
one also to ask just what changes that transition may have involved. But
while these questions are obvious ones to pose, and have been asked regularly for well over a century, this doesn't mean they have been answered
to everyone's or even to anyone's satisfaction. "Human" means different
things to different people, and even to the same person in different situations. For example, I am happy to use the term "human evolution" when
referring to the entire history of humankind, back to its common ancestor with today's great apes. In that context, the term "human" is pretty
much equivalent to "hominid." But does this really imply that all hominids are "human"? I, for one, would be very reluctant to use this term to
describe any of the bipedal ape species that populate the first few million
years of that history; and indeed I only find anything I might want to
call "fully human" at the very uppermost tip of the human evolutionary
tree. But that's just an opinion, and there's plenty of room for legitimate
disagreement here—there is certainly no official, or even generally accepted, definition of this elusive word "human." Remarkably, we have
hardly advanced in this respect since, some two and a half centuries ago,
Linnaeus's almost exact contemporary Samuel Johnson defined *human*
in his great English Dictionary as "having the qualities of a man," and
man as "a human being." Still, even though paleoanthropologists are a
famously argumentative breed, it's probably fair to say that most of us
broadly concur that the first creatures we can in some meaningful sense
call "humans" are the most ancient representatives in the fossil record
of our own genus *Homo*.

Unfortunately this consensus in principle hardly clarifies matters
much in practice. For there is no agreement on what the "qualities
of a man" actually are, even in the relatively simple terms that necessarily apply to fossil forms known only from bones and teeth. As a
result, there is a lot of confusion about exactly which fossils should
be placed in the genus *Homo*. To understand the current state of play,
we need to return to history for a moment. As we saw in chapter 4,
back in the 1960s Louis Leakey and his colleagues extended the definition of the genus *Homo* back beyond *Homo erectus* by whatever
it took to include in it the gracile "handy man" fossils from those
1.8-million-year-old rocks at the bottom of Olduvai Gorge. Although

the partial lower jaw that Leakey and his associates deemed the holotype of *Homo habilis* didn't look vastly different from its counterparts
among the gracile *Australopithecus* fossils from South Africa, Leakey
felt that some fragments of braincase indicated a brain a bit bigger
than was typical of the latter (though it was still smallish, at under 700
cc). In addition, the lower jaw was putatively associated with a partial
foot of what had clearly been an upright biped, with an in-line great
toe and nicely sprung arches. At the time there was nothing remotely
comparable to this foot so far back in the hominid fossil record, and
its features fit nicely with Leakey's long-established predilection for the
notion that the roots of our genus lay way back in time—just as the
crude stone tools found in the same sediments matched perfectly with
his attraction to the idea of "Man the Toolmaker." So it was that the
morphological concept of the genus *Homo* became stretched to include
some very ancient morphologies indeed.

It took a few years for paleoanthropologists to become comfortable
with the idea of embracing the rather archaic-looking Olduvai hominid
within our own genus. But once they came around to the rather bizarre
notion that the genus *Homo* could somehow accommodate a range of
morphologies extending all the way from modern *Homo sapiens* to the
ancient Tanzanian fossils, the way was open for them to start including
in *Homo habilis* a motley assortment of specimens from other African
sites. The process started in 1972, when the toothless 1.9-million-year-
old cranium KNM-ER 1470 from east Turkana was discovered, and
shortly thereafter hailed as the best-preserved *Homo habilis* skull yet.
Its allocation to *Homo* was made largely on the basis of an impressive
estimated brain volume of about 800 cc (later reduced to 750 cc); but
as noted the specimen is rather poorly preserved, and it is still hard to
know quite what kind of hominid it represents. The discovery of 1470
was followed by a flowering of other hominid fossil finds in eastern Africa, and diverse cranial and postcranial specimens, from Olduvai Gorge
and East Turkana—and as far away as South Africa—were subsequently
shoehorned into *Homo habilis*. As each of these fossils was engulfed, the
plasticity of the genus *Homo* appeared even greater.

Ironically, even before the extreme untidiness of this assemblage
of fossils became too blatant to ignore, our old friend 1470, the very

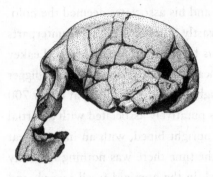
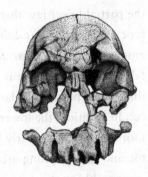

The partial skull KNM-ER 1470, from East Turkana, Kenya. Some 1.9 million years old, this individual had boasted a brain of about 750 ml in volume, larger than is typical for australopiths; and its discovery convinced many paleoanthropologists that Homo habilis *was indeed a real species. Drawing by Don McGranaghan.*

hominid that had convinced most paleoanthropologists that *Homo habilis* was a reality to be accepted, became the standard-bearer of a new name. In the mid-1980s a Russian paleoanthropologist renamed this fossil *Pithecanthropus rudolfensis* (oddly using Eugene Dubois's ancient genus name, rather than the universally accepted *Homo*). Within a few years, other paleoanthropologists started to pick up on the species *Homo rudolfensis;* and, in parallel with the ballooning of *Homo habilis,* this second ancient species of *Homo* acquired new exemplars in Kenya and even as far afield as Malawi. Some of these fossils are as old as 2.5 million years, but most date from around 2.0 million years ago or a little less, and all are pretty fragmentary.

The critical time span between two and two and a half million years ago also coincided with that of a number of finds in eastern Africa that their discoverers, presumably a bit worried about the increasing disarray of *Homo habilis* (and indeed, of *Homo rudolfensis*), diplomatically preferred to allocate simply to "early *Homo.*" Before the remarkable finding of those cut-marked bone fragments at Dikika, the 2.5-million-year date of the oldest of these fossils coincided pretty closely with the earliest evidence for the use of stone tools, and this was something that fed back powerfully into both the "early *Homo*" and the "Man the Toolmaker" ideas. Still, putting any of these fossils in our genus is a bit of a stretch, based solely on their preserved anatomy; and, as new evidence accumulates, this coincidence is beginning to look less like a red herring than as the basis for a self-fulfilling prophecy that sent paleoanthropologists up a blind alley.

Fortunately, we don't have to wait long after the two-million-year point to start finding fossils that really do merit inclusion in *Homo* because of all the features they specifically share with us. We'll discuss them in a moment; but first I should point out that, unsettlingly, it is hard to know right now where those strikingly new and innovative fossil relatives came from. There is little to connect them directly to any of the "archaic *Homo*" fossils we've just been talking about; and while we know a large array of australopiths—and there can be little doubt that ultimately it was one branch of these early bipedal apes that gave rise to *Homo*—it is really hard to pinpoint where among these diverse creatures the origin of our genus lay. To put the situation in a nutshell, there is not one fossil among all those known in the period before about two million years ago that presents itself as a compelling candidate for the position of direct progenitor of the new hominids to come. All we can say right now is that the period between about 2.5 and two million years ago was clearly a time of continuing evolutionary ferment among members of the hominid family. The ongoing experimentation with the hominid potential that went on in this period is expressed in an intriguing diversity among the fossil hominids we know; but, to add to the uncertainty, it is a diversity that we still glimpse only dimly.

This uncertainty is partly due to the fragmentary nature of the evidence, but there is a good chance the glimpse is also dim because of a general reluctance among paleoanthropologists to accept even in principle that such diversity is indeed out there. One reason for this is that it is difficult to make sense of the abundant but frustratingly incomplete evidence that we have at our disposal. Sorting out species structure within a sample of fossils is the most basic of tasks a paleontologist takes on, but even at the best of times it is also often one of the hardest. The simplest default hypothesis at which you can arrive when you're poring over a table covered in fossil fragments is that everything you're looking at belongs to the same variable species. As such, you don't have to decide just where any possible demarcations lie. But this is only one factor; and to a large extent the reluctance to perceive diversity also stems from an underlying expectation about evolutionary pattern. It will require just a little bit more history to understand why, in recent decades, paleoanthropologists have tended to take such an extraordinarily inclusive approach to membership in our genus.

In the half century preceding World War II, paleoanthropology was the province largely of human anatomists, scientists whose training was in the minutiae of human physical variation. They were not forced to confront the riotous diversity of species in nature that other natural historians had to contend with. One byproduct of this insular history was that few paleoanthropologists of the period had much if any training either in evolutionary process, or in the procedures and requirements that should underpin the naming of new species. This led to the liberal description of new hominid genera and species, almost as if each new fossil that showed up needed to be baptized with its own genus and species name, much as Western humans individually receive family and given names. By the time World War II rolled around there were at least 15 hominid genus names commonly in use, and countless species—all for a fossil record that was then of modest size.

In the long run this was bound to be a pretty untenable situation. And it was particularly untenable at a time when a movement that became known as the Evolutionary Synthesis was taking firm hold in most areas of evolutionary biology. The Synthesis was a meeting of previously disparate minds in the fields of genetics, systematics and paleontology, each of which had previously itself harbored multiple versions of the evolutionary process. On the one hand, the Synthesis emphasized the importance of variation *within* populations and species of living creatures, and on the other, it preached the basic continuities of the evolutionary process. The Synthesis allowed for the splitting of evolutionary lineages of organisms (without which we would never have achieved the luxuriant diversity of Nature that we see today). But at the same time it stressed that evolutionary change was the result of slow alterations in the frequencies of genes within established lineages, under the guiding hand of natural selection. Species were hence viewed principally as arbitrary segments of ever-modifying lineages: as transitory units, highly variable at any one point in time. And they were thus expected to slowly evolve themselves out of existence. So compelling was this gradualist message that, between the late 1920s and the mid-1940s, the Synthesis became the central paradigm of evolutionary biology in the English-speaking world. Virtually the last holdout, as a result of its peculiar history, was paleoanthropology. But not for long.

Perhaps the most influential architect of the Evolutionary Synthesis was the geneticist Theodosius Dobzhansky, who was declaring as early as 1944 that based on the fossil evidence there had never been more than one (highly variable) hominid species at any one point in time. At an influential conference held at Long Island's Cold Spring Harbor Laboratory in 1950, Dobzhansky was joined by his ornithologist colleague Ernst Mayr, who took this proposition even farther. Mayr argued that culture broadened the human ecological niche to such a degree that, even in principle, there *could* only ever have been one human species at a time. And it's worth remembering at this point that Mayr's notion is intuitively a very attractive proposition to members of a storytelling species that also happens to be the only hominid in the world today. It is somehow inherently appealing to us to believe that uncovering the story of human evolution should involve projecting this one species back into the past: to think that humanity has, like the hero of some ancient epic poem, struggled single-mindedly from primitiveness to its present peak of perfection.

Although he'd probably never seen a hominid fossil in his life, Mayr then took all of the many genera that were cluttering the hominid fossil record, and reduced them to *one:* the genus *Homo.* What's more, he reduced the species involved to a mere three. These formed a single succession: *Homo transvaalensis* (the australopiths) gave rise to the middle stage we call *Homo erectus,* which ultimately transformed into *Homo sapiens* (which included the Neanderthals). Mayr's Cold Spring Harbor declaration hit paleoanthropology like a bombshell. Before long, even he was forced by ongoing discoveries of robust australopiths to admit that there had indeed been at least one side branch from the mainstream of hominid evolution. But Mayr's reductionist view of the human fossil record still held paleoanthropology in thrall for the next several decades. Perhaps because paleoanthropologists had never really paid much attention to evolutionary theory before, paleoanthropology found itself suddenly dominated by the Synthesis. Indeed, so thoroughly was the field traumatized by Mayr's remonstrations that, all through the 1950s and well into the 1960s, many paleoanthropologists hardly dared to use zoological names at all, preferring to refer to individual fossils by the names of the sites they came from. That way, they couldn't be accused by their colleagues of being biologically naïve.

Once the trauma had worn off to the point that they felt comfortable with zoological names again, paleoanthropologists lapsed into a style of taxonomic inclusiveness. The attitude seemed to be, if we have to use zoological names, let's use as few as possible. And even though the burgeoning fossil record has since made it impossible to ignore the fact that there was indeed a lot of hominid diversity out there, most paleoanthropologists still hew to the rather minimalist mindset that so thoroughly dominated in the days when most of today's leading practitioners were trained. Of course, paleoanthropologists aren't fools, and nobody denies any more that the human evolutionary tree looks more like a forking bush than a slender sunflower. Even more significantly, a lot of different hominid species are now widely accepted, as the illustration in chapter 2 shows. Nonetheless, despite wide recognition that there is a great deal more to the evolutionary process than simple lineage modification under natural selection, the gradualist mindset still lingers in a residual reluctance among paleoanthropologists to recognize more branches in that tree than absolutely necessary. Perhaps once this reluctance has faded a bit farther we will be able to look more realistically at the diversity of the "early *Homo*" fossils, and to detect where within that assemblage the roots of our genus actually lie.

We haven't yet got very far with this process, but an important first step was made in 1999 by the English paleoanthropologists Bernard Wood and Mark Collard. These scientists looked at the criteria their colleagues had used to place various very early hominid fossils into the genus *Homo*, and they rapidly determined that those criteria were deficient. Instead of starting with the mass of fossils that had been allocated to our genus, and rationalizing why they should all be classified in *Homo* as Louis Leakey and his successors had done, Wood and Collard began at the other end—with the defining species, *Homo sapiens*—and worked outward from there. Starting from scratch in this way, they concluded that any member of a morphologically coherent genus *Homo* had to conform to a set of criteria (including body size and form, reduced jaws and teeth, and lengthened developmental schedule) that excluded all the australopiths. What's more, these criteria also excluded all of the fossils variously allocated to *Homo habilis, Homo rudolfensis,* and "early *Homo.*"

Unfortunately, Wood and Collard also recommended that the expelled fossils should be rehoused within *Australopithecus,* something that made this genus even untidier than before. But that situation was alleviated to some extent when Meave Leakey and colleagues suggested a couple of years later that 1470 (and, by extension, *Homo rudolfensis* as a whole) should be assigned to their new genus *Kenyanthropus,* based on facial similarities to that equally imponderable skull from west Turkana. To create not only a new species but an entire new genus in this way was a brave move on the part of these scientists, especially given that their type specimen was far from wonderfully preserved; but it was a very necessary step, and with any luck it heralds a more realistic approach to hominid taxonomy by future researchers. Meanwhile, the expulsion from *Homo* of the rag-bag of fossils we've just been looking at makes for a much tidier notion of our own genus—although it still embraces a long time and quite a wide variety of morphologies.

FULL-TIME BIPEDS

When, in 1894, Eugene Dubois described the ancient hominid *Pithecanthropus* (now *Homo*) *erectus* from the site of Trinil, in Java, he knew it was old: part of a fossil fauna that contained not only many species but also many genera of animals that are now extinct. As for the hominid itself, he only had a couple of teeth, a skullcap, and some thigh bones that looked remarkably human to go on. Indeed, paleoanthropologists still debate whether the leg bones and the much more primitive-looking skullcap are properly associated. The skullcap was long and low, and had contained a brain of about 950 cc in volume. Its shape reminded many of the braincases of the later Neanderthals, the only other extinct hominids known at the time, though it was much smaller. The Neanderthals had, on average, brains as big as or maybe even bigger than ours (which have a mean value of around 1350 cc). In contrast, the skullcap was strikingly different from that of modern humans, with a strong brow ridge overhanging the (missing) eye sockets at the front, and a distinct angulation at the back. The leg bones, though, were very humanlike and firmly indicative of upright posture, which is why Dubois named this species as he did.

Advances in dating have allowed us to determine that the Trinil fossil is between about a million and 700 thousand years old, and subsequent discoveries elsewhere in Java have revealed that Trinil *Homo erectus* is part of an endemic hominid group (which included the famous Peking Man) that flourished in its eastern Asian redoubt from perhaps as much as 1.8 million years ago until as recently as 40 thousand. Although there is a good bit of variation among the fossil specimens concerned, it seems reasonable to embrace them all within *Homo erectus*. All share regional characters that clearly differentiate them from African and European hominids in the same time range.

Nonetheless, Ernst Mayr had insisted that *Homo erectus* was simply the middle stage in the evolving lineage that led from the australopiths to *Homo sapiens*, and many paleoanthropologists continue to agree with him. As a result, a motley assortment of fossils in the 1.9- to 0.4-million-year range has subsequently been attributed to this species, largely based on their "intermediate" age, rather than on what they actually look like. And while by now it's generally agreed that there are no *Homo erectus* fossils known from Europe, many scientists still like to refer to a group of fossils they call "early African *Homo erectus*." This is, however, to push the notion of *Homo erectus* beyond reasonable biological limits; and a preferable name for the early African forms is *Homo ergaster* ("work man," in rather Engels-like acknowledgment of the stone tools it made), a name bestowed on a 1.5-million-year-old mandible from East Turkana in 1975. If the truth be told, even the fossils sequestered away in *Homo ergaster* make up a pretty varied bunch. But its members do seem to belong to a fairly coherent larger group. Until the details are sorted out, the species *Homo ergaster* provides a fairly convincing umbrella for all of them.

THE TURKANA BOY

Until the mid-1980s, the iconic *Homo ergaster* specimen was a cranium known as KNM-ER 3733, which was discovered in 1.8-million-year-old sediments to the east of Lake Turkana in 1975. Ancient as it is, it does not in the least resemble anything we know from earlier times. Although its face sits boldly at the front of a weakly inflated cranial vault, it does not strongly project like that of an ape. It would, however, have possessed a somewhat protruding nose, and this is a striking departure from the flattened mid-face that characterizes the living apes and was also present in the australopiths. Its cranial vault had contained a brain of about 850 cc, well in excess of the brain size estimated for 1470, and not far shy of the much younger specimen from Trinil. Altogether, here for the first time was a hominid skull that anticipated what was to come, rather than harking back to the past. This was something that deserved consideration as a member of the genus *Homo*. The 3733 cranium lacked all teeth but one, but in conjunction with other partial skulls and teeth from East Turkana it established that by 1.8 million years ago some east African hominids had reached an entirely new plateau—what many paleoanthropologists would call a new "grade"—that is also exemplified by *Homo erectus*.

Quite how distinctive that new grade was did not become fully evident until 1984, when fieldwork on the west side of Lake Turkana turned up most of the skeleton of an adolescent male known technically as KNM-WT 15000, but more widely familiar as the "Turkana Boy." Before this discovery, various hominid postcranial bones had been recovered to the east of the lake, but apart from one partial skeleton that was riddled with pathologies, all such finds had been of separate elements—and there was no way to know for sure with what kind of hominid they were associated. Here, though, was the almost entire skeleton of an unfortunate individual who had died prematurely, face down in lakeside swampy mud, some 1.6 million years ago. Fortunately for us, his remains had been covered by soft protective sediments before they could attract the attention of scavengers. The result was a bonanza to paleoanthropologists. Because now, for the first time, fossils were available to them that showed just how a single *Homo ergaster* individual had actually been built.

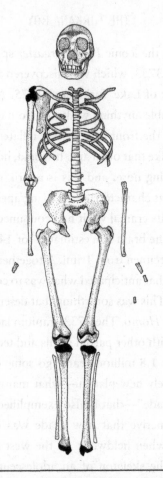

Skeleton of KNM-WT 15000, the "Turkana Boy," from Nariokotome, northern Kenya. About 1.6 million years old, this spectacular specimen is the only reasonably complete skeleton we have of a Homo ergaster *from East Africa, and though its brain was only of modest size it shows basically modern body proportions. Drawing by Don McGranaghan.*

It is a minor inconvenience that the Turkana Boy died before achieving maturity, complicating the task of reconstructing just what an adult *Homo ergaster* would have been like in life. Modern human children grow and mature very slowly compared to young apes (and australopiths), and they undergo an "adolescent growth spurt" beginning around the developmental stage at which the Turkana Boy died. It is reckoned that the Boy had stood about five feet three inches tall when he perished, and that if he had been poised to develop on a modern human schedule he would have stood about six feet one inch tall on achieving maturity. Tall, slim, and weighing maybe 150 pounds, in life he would have been a far cry from his small-bodied and stocky bipedal ape predecessors.

But there is also a major scientific advantage to his immaturity: we are able to see that the Boy had not developed as we do. Although his teeth had erupted and his bones had knit to about the degree you see in a modern 12-year-old, the painstaking process of counting the growth increments in his teeth under powerful microscopy indicated that he had actually lived for only about eight years. Evidently, his developmental schedule had been fast; and, although it was already modified in our direction, it had resembled that of apes more closely than it did that of modern humans. This in turn implies that, when he died, the Boy had already completed most of his growth. As a result, it's looking improbable that, even if he had lived a lot longer, he would ever have come close to hitting the six-foot mark.

Still, most modern humans don't hit that mark either; and development aside, what is most remarkable about the Boy is that his skeleton presents a striking contrast to that of Lucy and other bipedal apes. The Boy was tall, with long legs that contributed importantly to basic body proportions that are close to our own. Some echoes of the past remain; but in most essentials we see a creature not too dissimilar from ourselves, at least below the neck. Here, at last, is a hominid adapted to striding out across the open savanna, far from the shelter of trees. Gone are the "have your cake and eat it" ambiguities of the bipedal ape skeleton. The Boy's body is that of an obligate biped, rather than simply a facultative one: it is the body of a creature that was committed to upright bipedality as a way of life, rather than one that simply had this way of getting around as an option.

To put the situation another way, the Boy and others like him had adopted the savanna as their home. By 1.6 million years ago, grasslands had already become widespread in Africa, although those open, Serengeti-style savannas where the view goes on forever were still several hundred thousand years in the future. The environments through which Homo ergaster moved still largely resembled the mosaic of the past, with larger or smaller patches of grassland interspersed with clumped or scattered trees, true forest in hollows and along watercourses, and swamplands along lake margins such as the one on which the Turkana Boy died. But the new body form certainly reflects—or permitted, or even mandated—a novel way of exploiting the environment, with an emphasis on the resources available in the more open areas.

The definitive abandonment of the trees is reflected throughout the Turkana Boy's skeleton. For example, the extravagantly broad pelvis of Lucy had narrowed, apparently in concert with a lengthening of the leg. For while Lucy had needed a wide pelvis whose horizontal rotation could counteract excessive dropping of her center of gravity as she swung each leg forward, the Boy's long legs provided an alternate means of achieving the same thing. Compared to ours the Boy's arms were longish, but they were a far cry from those of apes. At the upper end they fit into a shoulder socket that faced out, like yours and mine, rather than up like that of an ape. But it also faced a bit more forward than ours does, and this has prompted speculation that the Boy's throwing capacities would have been limited. Sadly, the skeleton preserves few hand or foot bones; but the fact that a hominid like the Turkana Boy was striding around the Turkana Basin some 1.5 million years ago is confirmed by the recent discovery of large hominid footprints on the lake's eastern shore. These prints demonstrate both a long stride and a basically modern foot anatomy.

RADICAL CHANGE

The newly evolved body form represents a giant step along the road to becoming fully human; and, however exactly it was achieved, it is entirely unanticipated in the fossil record we have to hand. For, as I've already intimated, there is nothing in that record that we can regard as a convincing intermediate between any australopith or "early *Homo*," and the Turkana Boy. Based on the available evidence, then, the Boy does not fit at all comfortably into the expectation, derived from the Synthesis, that innovations should have appeared gradually in the hominid lineage. This discordance is hardly unprecedented: back in the mid-nineteenth century even Charles Darwin and his (otherwise) doughty defender Thomas Henry Huxley were already in deep disagreement over whether or not "Nature makes jumps." Darwin focused on slow, incremental change, while Huxley was worried by the many discontinuities he saw in the fossil record—and in nature in general—that were inconsistent with this pattern. Darwin's favored mechanism of natural selection provided a persuasive mechanism for gradual change, but Huxley's reservations were based on compelling evidence. Fortunately, recent advances in molecular genetics are finally helping us to understand what must actually have happened in the origin of human body form, as well as in a host of other apparent natural discontinuities.

All physical innovations originate in mutations—spontaneous changes—that constantly occur in our DNA, the hereditary molecule that makes up the chromosomes that reside in the nuclei of our cells (including the sex cells that combine at fertilization to produce each new individual). Particular stretches of the long DNA molecule equate to individual "genes," the units of heredity that were already envisaged long before the organization of DNA was understood. It used to be thought that the genes lay along the chromosomes pretty much like beads on a string, and that each gene was coded for the production of one of the protein molecules that serve as building blocks for the multifarious kinds of tissues that make up the developing body. This tidy image fit nicely with the gradualist predictions of the Synthesis. According to this interpretation, natural selection was simply a matter of eliminating most mutations while promoting others; and evolutionary change summed out to the gradual accumulation of favorable mutations within lineages, as one bead was incrementally replaced by another. However, since the basic structure of DNA was decoded in the early 1950s, we have learned that things are not that simple at all.

This brings us back to a subject I've already briefly introduced. It has long been known that most protein-coding genes act to determine more than one physical characteristic, and that most physical features are determined by several genes. But it was widely assumed that there must have been a general relationship between the number of genes and the complexity of the organism, and the recent discovery that humans, with their billions of cells, have only around 23,000 protein-coding genes—about the same number as a tiny nematode worm with only 1,000 cells—

came as quite a shock. What's more, the protein-coding genes turned out to make up only about two percent of the whole genome, as the totality of the DNA in our cells is called. How could so few genes govern the development of an organism as intricate and complicated as a human being? And what was the rest of the "junk" DNA doing?

The answers to the two questions are closely related. Some ingenious recent investigations have shown both that the effects of a coding gene depend largely on when and how long it is active in development, and that part of the "junk" DNA is significantly involved in switching protein-coding genes on and off during that process. It also turns out that a coding gene's effects depend on how active it is during the time it's enabled by those "switch" genes, and that yet other stretches of "regulatory" DNA govern the vigor with which the coding genes are expressed in the development of the tissues. What is more, differences in the expression of the same basic gene may have huge consequences for the phenotype (the observed features of the individual). It has turned out, for example, that the genes governing the development of chimpanzee and human brains differ much more in expression than they do in structure. Compared to their counterparts in chimpanzees, some 200 genes involved in human brain development were found in one study to be upregulated," and thus much more active; interestingly, this difference was lower in the brain than in other tissues of the body such as testes, heart, and liver, suggesting that the brain is under peculiar constraints as far as change itself is concerned.

The system of DNA-governing-DNA is the key to how so few coding genes can do so much work. This division of labor also explains why the genomes of *all* organisms have turned out to be strikingly similar. As recently as two or three decades ago, geneticists assumed that the genes determining what a fly and a human looked like would be completely different; but we have learned since that both animals are playing to a remarkable extent with the same genetic deck. When you consider that flies and humans share a common ancestor (albeit one that lived more than half a billion years ago) this is less surprising in retrospect than it seemed at the time it was discovered. But it is nonetheless amazing that such hugely dissimilar organisms can have about a third of the same basic genes in common. Of course, those genes vary structurally among species—which is why they are useful to systematists trying to figure out the relationships among the organisms they study. But, especially among close relatives, the difference in the phenotypic results produced by the coding genes may be due as much to the combinations in which they act, and to variations in their timing and expression, as to their basic structures.

This fact provides the key to understanding how Nature occasionally makes those leaps that so concerned Thomas Huxley. In the 1940s, the geneticist Richard Goldschmidt found himself roundly excoriated for suggesting that subtle genetic modifications might produce large phenotypic differences; after all, this was the heyday of the Synthesis, and Goldschmidt's choice of the term "hopeful monster" to characterize the transformed organism was perhaps unfortunate. Now, however, it is well established that structurally small genetic changes can produce new adaptive types, and that such innovations can at least occasionally be evolutionarily advantageous. The classic example is furnished by the stickleback, a small fish that boasts sharp spines, derived from the pelvic skeleton, that make it hard for predators to swallow. Some bottom-living sticklebacks, however, find those spines a distinct disadvantage—they can, for example, be grabbed by dragonfly larvae anxious to feed on their possessors. As a result the bottom-livers have lost the spines, apparently quite rapidly and recently. The modification is hardly trivial, involving as it does the elimination of an important part of a complex structure. But this major physical alteration has recently been shown to have occurred in the absence of *any* change in the coding genome. Instead, a small stretch of regulatory DNA has been deleted. This has left the basic gene intact to do its essential task, but it has eradicated the development of spines by reducing its activity in a specific area of the body. A tiny change in the genome has produced major phenotypic results. Most changes on this scale will actually be disadvantageous, and mechanisms of this kind certainly do not exclude the importance in stickleback evolution of mutations with smaller, more localized, effects. However, in the bottom-feeding sticklebacks' case this particular change just happened to have been an advantageous one, and clearly it spread very rapidly.

Perhaps the Turkana Boy's radically new bodily conformation can be attributed to a genetic event of similar kind. A minor mutation had occurred in the Boy's lineage that, through altering gene timing and expression, had radically changed its possessor's morphology—and had, entirely accidentally, opened new adaptive avenues to them. So maybe we don't need to ask ourselves why there are no harbingers of the Boy's radically new body morphology in the known fossil record. Perhaps there simply *weren't* any such intermediates—or at least none that we could reasonably expect to find on the coarse time scale that the fossil record represents. Something routine and unremarkable on the genomic level had occurred among the Boy's precursors; and it just happened to change the course of hominid history.

Further studies have shown that the Turkana Boy's rapid developmental timetable was not unusual. This somewhat apelike rapid growth seems to have been typical for hominids like *Homo ergaster* and *Homo erectus,* and results very similar to those on the Boy's teeth have been obtained from dental analyses of *Homo erectus* specimens from Java. Put together, these observations have substantial implications for what hominids of what paleoanthropologists like to call this "grade" (general kind) and time period may have been like in life. This is particularly true for the indications we have about brain growth, starting very early in the developmental process. Apes grow up much more quickly than humans do, and they go straight from juvenile to adult, omitting a prolonged adolescent developmental phase. Perhaps surprisingly, though, their gestation length is about the same as ours—though the process itself is subtly different. The main difference in the prenatal period is that, in the last trimester of pregnancy, human beings dedicate much more energy to the development of the brain than apes do. The result is that a human neonate already has a larger brain than its ape equivalent. And while this may be perfectly fine on its own, there is a pretty severe limit to the size of any head that can pass without difficulty through the fairly rigid pelvic birth canal.

Today, *Homo sapiens* is pushing uncomfortably hard against that limit, as witness to the distressingly high number of deaths in childbirth in the absence of modern medical supervision. (Somewhere in the world, a woman dies in this appalling way about every 90 seconds.) It has been suggested that, with their newly narrow pelvis, even a modest increase in the head size of the neonates they gave birth to might have required *Homo ergaster* mothers to have assistance during the process: a situation that would dictate some sort of midwifery. This idea has its own implications for social and cognitive complexity, and remains speculative. What is not in doubt, though, is that obstetrical requirements inevitably restrict the amount of brain enlargement that can take place before birth, meaning that to attain their large adult brains, modern humans subsequently need to divert large amounts of energy to brain development over sustained periods of time. The upshot is that an ape is born with as much as 40 percent of the brain volume it will have as an adult, while, despite that accelerated prenatal expansion, the equivalent percentage for humans is only about 25. Hence, in contrast to the slowdown in brain growth after birth that occurs in apes and other mammals, the human brain, exceptionally, keeps on expanding at fetal rates for at least the first year of life. So, while by the end of that first year an ape's brain has already attained 80 percent of its adult size, this compares to only 50 percent in human infants—whose brains will necessarily continue growing much longer, reaching adult size around the age of seven.

The Turkana Boy died at a stage of maturity when his 880 cc brain would already have been very close to adult size, so his fossil remains can't tell us much about his early brain development. But other evidence confirms that individuals of the *Homo ergaster/Homo erectus* grade conformed much more closely to the ape pattern than to the human one in brain development, as well as in other aspects of growth. Thus a recent study of a juvenile *Homo erectus* from Java, who died perhaps as much as 1.8 million years ago at about one year of age, showed that even at this tender age its fast-developing brain was already around 72 to 84 percent of the average adult size for the species.

This accelerated schedule of brain development has implications both for the mental complexity of those ancient members of our genus, and for the kinds of lives they lived. Modern humans are "secondarily altricial," meaning that infants of our species are produced in relatively small numbers but are helpless or extremely dependent on their parents for an extended period—a period that, in our case, is associated with a great deal of complex learning and transmission of social skills, including the acquisition of language. It may also be associated with an increase in the complexity of the social apparatus for bringing up infants, with more generations becoming involved in the process. Great apes become sexually mature, and conclude their key learning period, at about seven years of age. In contrast, modern human beings take almost twice as long to become sexually mature, and considerably longer still to complete their physical and emotional development. The fact that their brains are still immature and incapable of fully assessing risk is, for example, a major reason why teenage drivers have such an appalling accident rate. To be sure, the faster-developing apes are remarkably sophisticated beings, with highly nuanced societies and complex inter-individual relationships; but although they show certain rudiments of what we might in the broadest sense recognize as "culture"—the local, rather than species-wide, transmission of learned traditions—they are clearly not cultural in the highly complex human manner. Certainly, what any living human being needs to know to be an integrated member of society is hugely greater than anything an ape is required to master.

So where did the Turkana Boy and other members of the *Homo erectus* grade fit into the spectrum of developmental timetables? And how would this have affected or fed back into their cognition? Well, if we can take it at face value, the relatively fast maturation of these hominids strongly suggests that, for all their innovative physical similarities to us, on the cognitive side they were very different from today's *Homo sapiens*. They were, indeed, unique: they were neither bipedal apes nor modern humans; and though they had clearly progressed well beyond the ape stage, they led lives that were not mediated by mental equipment that was anything like our own. And while as adults they had brains that in absolute terms were considerably larger than those of the bipedal apes, those brains were not a whole lot bigger when compared to their greater body size. This is a key consideration, because the larger your body is, the more brain you need to control its basic motor and sensory functions.

Of course, the vast range of brain sizes found among cognitively normal modern humans suggests that *within* any species there is no close relationship between brain size and smarts. But *among* species the story is different. If you plot out the relationship between brain and body sizes among mammals in general, you find that there is indeed a strong relationship between the two variables. As body sizes increase, so do brain volumes (though typically in mammals brain volumes don't increase as fast as body size). One of the many notable features of *Homo sapiens* is, though, that our large brains place us way above the curve that describes this basic relationship. We have much bigger brains than you would predict for a mammal of our body size. But the Turkana Boy and his kin were less remarkable in this respect. They departed much less strikingly than we do from the typical primate brain–to–body size relationship, although they also diverged from the specifically great ape pattern of relatively smaller brains with increased body size. These early members of our genus may well have been the smartest mammals of their day; but they almost certainly perceived the world, and processed information about it, differently than we do. They were a long way from being simply junior-league versions of ourselves, and we should resist the temptation to view them that way.

Such resistance is particularly important when we look at the shape of the Turkana Boy's brain. Unlike most bones of the body, which are pre-formed in cartilage that is gradually converted to hard bone as the individual grows, the bone of the skull vault forms in membranes that are carried outward by the expanding brain inside. Most of the increase in the size of our brains relative to those of apes is accounted for by expansion of the cerebral cortex, the outer layer of the brain; and since large-brained hominids are cramming a lot of extra cortex into a relatively small space, this expansion has caused the cortex to become deeply folded and wrinkled over the course of human evolution, providing a larger surface area. The key thing here is that the major wrinkles outline what have traditionally been identified as major functional areas of the brain; and because of the intimate developmental relationship between the bone and the outside of the brain, the inside of the skull vault provides a record of these important demarcations. The brain itself doesn't preserve, but since it fits so closely into the inside of the braincase that contains it, an impression (or "endocast") of the inside of a fossil skull such as the Boy's can accurately represent what the organ it contained had looked like externally. There is, of course, a limit to what you can actually tell from this information, because how the brain is wired internally is critical to how it works; but nonetheless the external details can be informative.

One of the things that caught researchers' eyes early in studies of the Turkana Boy's endocast was that it prominently outlines a small region, called "Broca's area," that lies on the left frontal lobe of the cortex. The eponymous Paul Broca was a nineteenth-century French physician who noticed that patients with injuries to this particular area of the brain typically had difficulty speaking, even though they still readily comprehended speech. Clearly, here was a part of the brain (actually, two parts, since neuroanatomists now subdivide it on grounds of cellular structure) that was somehow involved in the production of speech, and it was among the first identifiable external brain areas to be implicated in a specific function. This was an important step in the recognition that specific areas of the brain—different clusters and types of neurons—are responsive in particular tasks. We don't think or respond to stimuli in a holistic way, with our whole brains. In a way this is disappointing, because it means that paleontologists can't take absolute or even relative brain sizes as proxies for anything very specific. But it makes everything much more interesting.

Perhaps the most significant advances since Broca's time in understanding how the brain works have been made possible by the development of techniques for imaging the activity going on in living brains while their possessors undertake various mental tasks. An important result of such real-time investigations has been the realization that most functions, including speech, are more widely distributed in the physical brain than simple superficial mapping would suggest by itself. Nonetheless, the identification of Broca's area in the Turkana Boy led to speculation that the Boy could have talked. But nothing is quite that simple, and Broca's area is now also known to be involved in a whole slew of memory and executive functions unrelated to language. Clearly, having one of the many features whose proper function is necessary for speech production cannot be taken as *prima facie* evidence of speech itself; and anyway, the possession of structures that might be related to the latent ability to produce speech is a long way from implying that these hominids had language as we know it.

Another aspect of the Boy's anatomy also argued strongly that they didn't possess linguistic skills. The spinal column not only supports the upper body but also conducts the spinal cord down from the brain, to control and receive information from the rest of the body via the network of nerves branching from it. The width of the canal through which the cord travels is pretty constant in most primates, including hominids; but in modern *Homo sapiens* (and, to be fair, in Neanderthals as well) it is unusually broad in the thoracic region, where the lungs sit. This additional breadth accommodates an increased volume of nervous tissue supplying the muscles of the thorax and abdominal wall, and it is suggested that the extra nerves are devoted to an enhanced control of breathing—a fine control that is, among other things, necessary to produce the subtle modulation of the sounds we use in speech. Interestingly, the Turkana Boy is an average primate in this regard. And it has accordingly been suggested that, regardless of the properties of his brain, he did not possess the peripheral ability to produce speech.

It has also been controversially proposed that there may have been something pathological in the narrow space for the Boy's spinal cord. That's as may be; but there are plenty of independent reasons to believe that however he and his kind communicated—and they undoubtedly had a sophisticated form of communication—they did not share information using language as it is familiar to us. For a start, modern articulate language is the ultimate symbolic activity, and we find nothing in the archaeological record associated with the *Homo ergaster/Homo erectus* group, at any point in its long tenure, that suggests any symbolic mental manipulation of the information received from the outside world. Indeed, sketchy though it is, the archaeological record left behind by these early members of the genus *Homo* is remarkable for the conspicuous *lack* of any such evidence. Perhaps most strikingly, the Turkana Boy and his kin made stone tools that were identical in concept to those that had been made at Gona almost a million years before. On the technological level, nothing significant had changed that we can detect over that whole vast period of time. The appearance of a radically new physical type had not resulted in—or from—any kind of technological innovation; and we have little material evidence to confirm that *Homo ergaster* had a substantively different lifestyle from its predecessors, even though the anatomical indicators prompt us to speculate that there must have been changes.

While it may seem counterintuitive that a new (and larger-brained) hominid should not have brought a new technology with it, this disconnect actually reflects a pattern already established among hominids: the very first toolmakers had, after all, been bipedal apes, not members of the genus *Homo*. This pattern set a template for future developments in that we cannot associate any later introduction of a new technology with the appearance of a new species of *Homo*. And this makes a lot of sense when you think about it, because ultimately a technology has to be invented by an individual—who has to belong to a pre-existing species. Innovations of all kinds must originate *within* species, if only because there is nowhere else for them to happen.

SIX

LIFE ON THE SAVANNA

The extraordinary skeleton of the Turkana Boy gives us a remarkable insight into his species *Homo ergaster:* a hominid that grew up fast but physically was like nothing we know from earlier in time, and a creature that was clearly at ease away from the ancestral forest. This radically different environmental setting made enormous new demands upon the young species, but it clearly did not initially respond by making technological adjustments: as far as we can tell, the very first *Homo ergaster* continued making the same kinds of tools that its more anatomically archaic predecessors had made. And in the absence of substantial evidence of technological change, we have to fall back on physical and other indirect indicators if we want to understand what was new in the life of *Homo ergaster.* But these indicators are highly suggestive, even though we are hard put to draw specific conclusions.

Although the Turkana Boy had a slender build, he was no weakling. In mechanical terms the shafts of the long bones of the limbs are basically hollow cylinders; and although the material of which they are made is hard and strong, it is not static. Instead, it remodels throughout life to resist the stresses placed on the limbs; and the varying thicknesses of the shaft walls reflect how high those stresses were in life, and how they were distributed. This is why fencers and tennis players have stouter bones in their dominant arms than in their passive ones, while astronauts' bones thin out after too much time in microgravity. One major respect in which the Boy's limb bones differed from ours is that, as in other early hominids, their shaft walls were much more robust than you see in humans today. This could indicate that in life the Boy was already immensely strong, and that he had also maintained a much higher level of activity than is typical of modern humans. Of course, our contemporary sedentary lifestyle is a very recent phenomenon; but even our ancient hunting-and-gathering *Homo sapiens* forerunners had relatively thin-walled long bones. Overall, since the Boy's time the thickness of the bone in the shafts of the limb elements has plummeted, implying that bodily strength has become a significantly less important factor in the hominid way of life.

The Boy's environment was not an easy one, and at least to begin with he and his fellows were out there on the still tree-studded African savanna without a significantly improved toolkit. There is every reason to believe that—as relatively poor climbers—they could not and did not depend on the trees for shelter to the extent their bipedal ape predecessors had. And in the more open areas they would have favored, there roved an array of predators as fearsome as those still lurking on the forest fringes. Mainly, but far from entirely, these were big cats of a much greater variety than is found in Africa today, and they were all ready to pounce on any unwary mammal they encountered. By our standards *Homo ergaster* individuals were strong; but they were nonetheless relatively defenseless, lacking big jaws and slashing canine teeth. How did they respond to this hazardous new environment? And how did they exploit it? There is no shortage of ideas; and although there is little evidence to substantiate any of them, a circumstantial case can be built.

One coherent scenario involves the notion that, with its modestly increased brain size, *Homo ergaster* needed a higher-quality diet than the varied but still plant-based one on which its predecessors had subsisted. This is because, although the benefits of a bigger brain seem self-evident to us *Homo sapiens,* the costs are at least equally evident. As I've already briefly noted, in metabolic terms the brain is among the most "expensive" tissues of the body. For, while the mass between our ears only accounts for some two percent of our body weight, it actually consumes something between 20 and 25 percent of all the energy we take in. This has major implications for the body's overall economy, including that of the digestive system. The broad abdomens of the Turkana Boy's australo-pith ancestors had almost certainly contained huge digestive systems, a feature that stands in stark contrast to modern humans. One of our most striking characteristics—almost as striking as our large brains—is that we have remarkably small internal organs for our body size. This was also true for the relatively narrow-hipped *Homo ergaster*, and it has important implications for the diet of the Turkana Boy and his fellows. For the internal organs are almost as "expensive" in energetic terms as the brain; and it has been powerfully argued that gut reduction in human evolution has not only been a necessary trade-off for brain expansion, but that at the same time it also exacerbated the need for a high-quality diet. Thus, although at the time of *Homo ergaster* hominids were only at the very beginning of their period of dizzying brain expansion, their reduced guts alone may have mandated dependence on high-yield foods.

Diorama in the American Museum of Natural History, showing two Homo ergaster *in northern Kenya, some 1.8 million years ago. It is left to the viewer to decide whether the hominids had scavenged or hunted the impala they are shown butchering. Figures by John Holmes. Photo: Denis Finnin.*

So where did the extra energy come from in the big-bodied and modestly brained *Homo ergaster?* One obvious answer is that these early hominids had turned their attention to the highest-quality diet available to them: animal proteins and fats. This resource was, after all, roaming the African savannas in enormous quantities: mammals of all sizes abounded in the newly adopted environment. At the same time, though, these tasty beasts also attracted a hugely diverse fauna of specialized carnivores, far more numerous than their counterparts occupying the continent today. In going after savanna grazers the hominids would have had not only to compete for their food with these professional predators, but they would also have had to protect themselves from them.

Perhaps less dangerous would have been fishing, and there are reasons for thinking that this activity may have been more important for *Homo ergaster* (and its successors) than the material evidence indicates. Aquatic animals are an important source of nutrients, such as omega–3 fatty acids, that are important for normal brain function. Limited quantities of these—enough, for example, to sustain a small ape brain—can be synthesized by the body. But the greater amounts demanded by an enlarged brain can only be supplied by the diet, and it has been suggested that ingestion of fish and other aquatic creatures may have been one precondition for the increase in hominid brain sizes over the past two million years or so. Many primates—particularly macaques—have been observed to obtain and eat aquatic invertebrates, and in one place orangutans have been observed fishing by hand. It would not have been difficult for early *Homo* to obtain fish in shrinking ponds and streams during the dry season, so it seems likely that they augmented their diet with such resources.

Whatever their origin, not only are animal products rather indigestible without treatment of some kind, but meat is also intrinsically pretty tough to get hold of. Potential prey animals are difficult to acquire because they hate to be eaten—they don't wait around like tubers or fruits to be dug up or picked by savvy foragers. They get out of the way, fast. This propensity would have posed a problem for any savanna newcomer bent on exploiting even small-bodied animals as a primary resource. Still, some researchers believe that, with a few behavioral innovations of the kind we wouldn't necessarily expect to find reflected in the material record, hominids could have effectively hunted larger mammals using the physical advantages offered by the new anatomy alone. They point to the fact that, although *Homo ergaster* would hardly have been fast compared to four-legged predators, its new slender hips and long legs would have made members of the species exemplary distance runners. In the heat of the day, the human ability to simply keep going would have allowed these lanky bipeds to single out, say, an antelope, and to keep chasing it until it fell from heat prostration.

Such a strategy would not only have been metabolically expensive, but it would also have necessitated the mental concentration to visually follow an animal to the horizon and, should it disappear from visual range, to track the quarry using spoor, broken branches, and other indirect signs. This mode of pursuit is used by African hunter-gatherers today (who wisely tend to walk or trot, rather than run, as running on soft surfaces has turned out to be as hazardous as running on hard ones), and it is made possible not only by sophisticated cognition of the hunters but by the physiological differences between them and their prey. Though faster than humans, most mammals do not have the capacity to shed the heat load acquired and generated during sustained activity in the tropic sun, except by pausing in the shade while it slowly dissipates, largely through panting. Hairless humans, on the other hand, constantly shed heat by sweating and radiation, allowing them to keep going when other animals drop from heatstroke.

It is impossible to know for sure whether *Homo ergaster* was indeed hairless and sweaty: even today, we retain our covering of body hair, though in such a reduced form as to be invisible in most places. But advocates of the notion that *Homo ergaster* was naked-skinned ingeniously point to some interesting studies of human lice. Most kinds of mammal only support one louse species; but humans have the luxury of sustaining two. One of these parasites lives in the hair of the head; the other inhabits the hair of the pubic region. A bit embarrassingly, while the human head louse is distinctive, the human pubic louse is a close relative of the form that inhabits gorillas, and is thought to have been acquired from this source. The head louse seems to be a survivor of the form that roamed all over the body of the human ancestor, while the pubic louse was acquired subsequent to the loss of body hair. Using the "molecular clock" (basically, the assumption that mutations in the DNA accumulate at a more or less constant rate), parasitologists have been able to estimate that the two kinds of lice parted ways some three to four million years ago; and this date range would logically indicate that body hair loss had occurred well before the time of the Turkana Boy, and perhaps even before that of Lucy.

While the parasite data may be a bit controversial, there is nonetheless general consensus that once modern body form had been achieved, luxuriant body hair was gone. Away from the trees, out in the tropic African sun, the physiological rules had changed; and because it is such a good bet that heat-shedding by sweating was the primary means of keeping the brain and body cool, the betting has to be that the Boy and his like were bare-skinned. What's more, in an environment of intense solar radiation, that skin would have been very dark. As northerners who have spent too long on a tropical beach know well, light skin is highly sensitive to ultraviolet radiation, and it's no coincidence that the highest skin cancer rates in the world today occur in Australia's sunny Queensland, where fair-skinned folk are wont to unwisely disport themselves in minimal clothing.

Perhaps, though, these behavioral speculations sound a little too human to add up to a convincing picture of *Homo ergaster,* as if our main behavioral characteristics had been established at that remote point in time, and that all that remained was for hominids to wait around another million and a half years for their brains to become bigger. What's more, the endurance-hunting scenario begs a number of important questions. Among these is whether *Homo ergaster* had the technology to carry around water—because while sweating may be an efficient way to lose heat, it is also a very effective way of using up the body's fluid supplies. Replenishing those fluids while chasing animals all over the landscape in the hot tropic sun would have required the constant availability of water, and we have no direct evidence that *Homo ergaster* possessed the technology to provide the containers needed to accomplish this. On the other hand, since the perishable stomachs or bladders of medium-to-large-bodied animals represented the only plausible materials to transport water, we would not expect to find evidence of their use preserved. And we can't take absence of evidence as evidence of absence. Beyond this, it is fair to point out that there is nothing we know or can reasonably infer about *Homo ergaster* cognition that would rule out the possibility that these creatures used simple containers. We know for instance that, long before the genus *Homo* came along, the earliest australopith stone toolmakers were already exhibiting foresight and planning in the course of their daily activities. At a certain level these early hominids understood the properties of hard materials; why not of some soft ones, too? Still, it is notable that, where the necessary landscape archaeology has been done, hominid activity sites in the Turkana Boy's time frame typically occur near places where water had been available; only later do we find evidence that hominids were venturing limitlessly across the countryside. All in all, the picture is frustratingly incomplete.

FIRE AND COOKING

However exactly it was that the requisite high-quality diet was acquired, hunting is still an energetically expensive activity. So, especially for a hominid with a small gut, it is imperative to get as much as possible out of the results of the hunt. As I've already briefly mentioned in connection with the probably carnivorous propensities of the bipedal apes, one way of doing this is by cooking the carcasses of your victims. Raw meat is pretty indigestible if you don't, like a lion or a hyena, have a digestive tract that is specialized for the task. Even after endless chewing, chimpanzees with their large stomachs and long intestines excrete lots of undigested bits of meat in their feces after the hunt. Primate digestive tracts just don't do a great job of extracting energy from raw animal sources. But cooking changes the game entirely, and brings with it a long list of virtues. Judicious cooking makes foods—of all kinds, not just meat—easier to chew, and easier to extract nutrients from. It kills toxins, makes foods edible for longer, and just plain improves flavors and textures. At whatever point it was introduced, cooking clearly made a huge difference to hominid life.

Nonetheless, whether this activity really was essential in allowing a creature like *Homo ergaster* to flourish still remains a speculative matter. This is not least because cooking presupposes the mastery of fire, and there is precious little direct evidence that *Homo ergaster* had achieved this. There are a couple of early indications that fires had burned at hominid sites in the *Homo ergaster* time frame, in the form of apparently singed bones some 1.8 million years old from Swartkrans in South Africa, and scorched clay balls about 1.4 million years old at the robust australopith site of Chesowanja in Kenya. But although these objects do seem to have burned at campfire temperatures, it is hard to see them as definitive evidence of controlled fire under hominid supervision. The earliest substantial evidence of fire *control* turns up only very much later: from an 800-thousand-year-old site in Israel, whence hearths containing thick layers of ash have been reported. You can, of course, argue that fire use is not always going to leave traces that will preserve over the long term, and that the African archaeological record we have in this period is sufficiently sketchy as to leave lots of room for doubt; and you would be right to do so.

At the other end of the timescale, it is actively argued that habitual fire use was a late acquisition in hominid history. But there can nonetheless be no doubt that fire control would have been a revolutionary innovation in hominid life; and it certainly seems odd that once it had been invented it would not have been widely adopted, in which case there should be more and better evidence of it. There are numerous sites at which you would expect to find hearths if there had been any—and don't. And several hundred thousand years passed after the well-documented Israeli occurrence before we begin to pick up further evidence of controlled fire in hearths, quite plausibly at first in opportunistic rather than habitual contexts.

While it is impossible to ignore the fact that the evidence for fire use by *Homo ergaster* is almost entirely circumstantial, the case for believing that food was cooked by these early relatives remains mildly compelling. Further, it is bolstered by other considerations, albeit also indirect. There can be no doubt whatever that the use of fire would have made life a great deal easier for *Homo ergaster* groups out there on the savanna, and it's been argued that fire control was the only thing that could have rendered this new lifestyle possible at all. For the very resource—the grazing animals—that may have lured *Homo ergaster* out on to the savanna in the first place also made it a dangerous place for the hominids, who certainly did not fit tidily into a dichotomy between predators and prey.

While they may have been predators, at least to some degree, they were slow and vulnerable targets as well. Indeed, one *Homo ergaster* frontal bone, from a site in Kenya, shows carnivore tooth-marks above one orbit that suggest a violent demise at the claws of a predator.

Early hominids at this stage were essentially amateur hunters, just breaking into the business and low on the learning curve. Indeed, for all the vaunted technological prowess that has made us today's top predator, we have not entirely put the vulnerabilities of our remote past behind us. Any jogger mauled by a mountain lion, or bow hunter chased up a tree by a bear, will assure you of that. Fire would have been an excellent way for *Homo ergaster* to discourage predators, especially to make up for the limited throwing abilities of that forwardly rotated arm joint. And if you are prepared to pile on a few more assumptions, the implications of fire use go well beyond this: some authorities have even gone so far as to suggest that many of the behavioral hallmarks of *Homo*, including its high degree of sociality and cooperativeness, stem from the closeness among group members that huddling for warmth and protection around a fire would have involved, in those early times as much as at present.

THE SOCIAL SETTING

There is no denying that fire has a unique symbolic as well as practical meaning to human beings today, and it's important to resist the resulting temptation to anthropomorphize. Still, while it may be reading a bit too much into what is already an elaborate succession of assumptions to suggest that the domestication of fire is directly responsible for our uniquely intense form of sociality, it is certainly true that modern humans are strikingly cooperative compared to other primates. Beyond simple cooperation, though, they also share an elaborate kind of sociality—known as "prosociality"—that seems to be unique. To put this at its most elementary, humans care at least to some extent about each other's welfare; and chimpanzees—as well as probably all of our other primate relatives—do not. Of course, mother-offspring bonds among chimpanzees can last a lifetime; and hunting and similarly complex activities sometimes involve extensive coordination among group members. What's more, chimpanzees have been observed to console victims of aggression, suggesting that they have some form of individual empathy. But such manifestations are different from the general concern for others that underwrites prosociality; and, in a large body of experimental studies, chimpanzees have come across—even to chimpophile researchers—as creatures that show a striking lack of regard for their fellows.

Researchers have tested this in captive settings. In one series of experiments, carried out on a number of different captive groups in different locations, chimpanzees were in various ways given the option of obtaining a food reward both for themselves and a neighbor, or just for themselves. For the chooser, the reward was the same in either case; but invariably the chimpanzees chose more or less randomly among the two options. On the basis of these tests, at least, the individual chimpanzee subjects seemed pretty consistently indifferent to the interests of others; and this stands in striking contrast to humans, who in psychological tests seem to be remarkably willing even to incur costs to help strangers.

The chimpanzee results may, of course, reflect cognitive limitations that are not directly related to sociality; but no matter exactly what those limitations might be, it seems highly likely that the vulnerable *Homo ergaster* had in some way transcended them. Almost certainly, in its simultaneously hazardous and productive new habitat, *Homo ergaster* could not have got by without some of the cognitive and social qualities that are so much a hallmark of its descendants. Sadly, we are unable to say much more than this with any confidence; but there are certain other kinds of inference we can draw about the lives those savanna pioneers led, and about the sorts of groups in which they may have lived.

We saw earlier that the australopiths, vulnerable small-bodied creatures hovering about the forest fringes, may well have lived in very large groups dictated by their high vulnerability to predation. But for the Turkana Boy and his kind the calculation was probably quite different. If, as likely, these hominids had indeed acquired cultural ways of controlling the threat of predation in their new environment, then the pressures for maintaining large group sizes would have been lessened. And with greater reliance on animal products in the diet, the constraints that apply to any professional predator would have begun to assume more importance in determining how the hominids lived. In any ecosystem prey individuals vastly outnumber their predators, for too many predators would eliminate the prey in short order, to everyone's disadvantage. If *Homo ergaster* was in the early stages of committing to an at least partially predatory lifestyle, then the advantage would have lain in reducing population density, and by extension group sizes, for the very good reason that the number of individuals who can be supported is determined by the sustainable availability of resources within the area that can be patrolled by a single group.

Range size would have been yet further limited by the mobility of females, who gave birth to helpless babies that demanded intense care for prolonged periods. Among primates in general, newly born offspring cling to their mothers' fur, and although the physiological demands of lactation are high, carrying clinging infants around is not a technical problem so long as there is only one, or perhaps two of them, at a time. But *Homo ergaster* mothers probably had no fur for babies to hold on to, and transporting slowly maturing infants around would have been a considerable chore. As far as offspring were concerned, more would not necessarily have been better. Indeed, among documented recent hunter-gatherer groups living in environments comparable to those of *Homo ergaster*, population control has been a much greater concern than fertility. San women living in the Kalahari of southern Africa typically breastfed their infants for up to four years, thereby maintaining high levels of the hormone prolactin that inhibits ovulation. There's little doubt that high levels of infant and juvenile mortality due to predation and other causes would have favored the maximum number of offspring an individual *Homo ergaster* mother could cope with, but this limitation in itself would have assured modest group sizes. Within those small groups, it is plain that females burdened with infants would have benefited from bonds with males who could have helped them provide for their offspring; but whether or not lasting male-female bonds were formed within the group, has to remain purely hypothetical.

The number of individuals—a dozen? twenty?—within a typical group of *Homo ergaster* also has to remain a guess, though it's clear that group sizes would have varied from place to place depending on the productivity of the local environment. Hominid groups would certainly have roved widely over large areas, perhaps splitting up into smaller foraging parties as circumstances dictated, occasionally encountering others of their kind, and availing themselves consistently of plant foods while obtaining animals when they could. Cut-marks made by ancient stone tools on butchered bones sometimes overlie carnivore bite-marks, which suggests that carcasses were sometimes scavenged—perhaps by power-scavenging. But in other cases butchered carcasses are entirely free of bite marks, indicating probable or at least possible hunting by hominids.

That groups of early *Homo* at least occasionally circulated around a central home base to which they frequently returned is also very much on the cards. At Olduvai Gorge, for example, evidence has been found for the processing of multiple carcasses in one spot within a single season. And at a Kenyan site called Kanjera it seems that, by some two million years ago, hominids were already regularly processing animal parts using tools made from stones derived from a variety of sources as much as 12 to 13 kilometers away. These findings have several intriguing implications, perhaps the most important among them being that, even before we have definitive fossil evidence that *Homo ergaster* was on the scene, hominids were already showing some key elements of later human behavior. But while the transport distances involved also suggest that by two million years ago hominids were already living highly energetic lives of the kind that analysis of the Turkana Boy's skeleton reveals, we can't be entirely sure exactly who those hominids were. With luck this issue will be clarified as that tantalizing assemblage of hominid fossils in the 2.5- to 2.0-million-year time frame is augmented and sorted out. Meanwhile, we can be confident that by the time the Boy himself came along some 1.6 million years ago, hominids were already living complex existences that anticipated major developments to come.

In any event, for all the sophistication of its immediate predecessors, it is clear that we cannot dismiss *Homo ergaster* as merely an advanced bipedal ape in a new kind of body. But at the same time, we can be fairly sure that this hominid's way of life exhibited a considerable degree of behavioral continuity with the past. This is hardly surprising in the larger context of hominid prehistory. For one thing, the knowledge that the first members of *Homo ergaster* wielded stone tools basically identical to those their predecessors had already been using many hundreds of thousand years before, gives us our first glimpse of yet another enduring hominid behavioral pattern: namely, the tendency to respond to fluctuating climatic and environmental circumstances by using existing toolkits in new ways, rather than by inventing new technologies. This is totally consistent with the fact that, from the beginning, hominids have always been ecological generalists. We have typically avoided the dangers of specialization by remaining unfailingly flexible in our responses to changing external conditions, in a sometimes dramatically fluctuating world. Hominids typically haven't so much *adapted* to change, as they have *accommodated* to it.

None of this means that gradually cumulative modifications couldn't have been made over the bipedal apes' long tenure. The lifeways of the australopiths had probably become more complex, and their exploitation of resources greatly refined, over the several million years of this ancient group's existence. Any such changes, though, were achieved in ways that are only indirectly reflected in the material record we have to hand so far. This is a pity, because it seems clear that the radically new *Homo ergaster* must have originated in some behaviorally sophisticated australopith isolate. And the chances are that, far from being propelled by circumstances, this new body build just happened to open up hugely advantageous new possibilities for its possessors in a new and expanding environmental setting. In keeping with this established pattern, the next technological leap most probably occurred, after a certain delay, within a population of *Homo ergaster* itself.

SEVEN

OUT OF AFRICA...
AND BACK

The human family was born in Africa, and many millions of years passed before we have any evidence that hominids had managed to escape the continent's confines. For a long time it was believed that the first hominid movement into Eurasia must have been facilitated by some dramatic acquisition such as an enlarged brain, or the acquisition of an improved technology. Now, though, things are looking a lot more murky; for the initial dispersal out of Africa appears to have occurred as early as 1.8 million years ago, or possibly more, and in a very archaic archaeological context.

The ruined medieval town of Dmanisi, lying in the Republic of Georgia between the Black and Caspian Seas, is the last place anyone would ever have thought of looking for early hominid fossils. Far from the layer-cake sediments and bare, sun-baked exposures of the Rift Valley, Dmanisi sits high atop a black basaltic bluff in a verdant landscape, dominating the confluence of two river valleys along which important ancient trade routes converged. Its setting alone makes it among the most spectacularly sited fossil localities anywhere. After a long and tumultuous history that ended in its comprehensive sacking by invading Turkoman forces in the fifteenth century, the formerly bustling town fell into decay. Unaltered for hundreds of years except by neglect, Dmanisi thus offered a wonderful opportunity to twentieth-century archaeologists seeking to understand more about life along a major branch of the great medieval Silk Road. Excavations in what remained of the dwellings revealed that the ancient townsfolk had dug circular grain-storage pits beneath their homes; and in 1983, entirely unexpectedly, fossil mammal remains were found in the walls of one of these. The town, it turned out, had been constructed on a thin veneer of sediments that blanketed the basalts below; and the first fossil to be found in these soft rocks, the tooth of a rhinoceros, turned out to belong to a species that was characteristic of the early Pleistocene. Suddenly, the earth below it had become even more interesting than the town itself.

Crude stone tools were found not long thereafter, and in 1991 the first Dmanisi hominid turned up: part of a lower jaw, with a beautifully preserved set of teeth. When it was announced in 1995 this specimen elicited broad comparison with *Homo erectus,* and it was felt to be most closely comparable with specimens from eastern Africa, with which the associated mammal fossils suggested it was roughly contemporaneous. This was later confirmed by dating of the underlying basalts to some 1.8 million years ago, in concert with geological analyses revealing that the uneroded basaltic surface had been covered and protected by the fossil-bearing sediments not long after its extrusion over the landscape. Shortly before the early age of Dmanisi was verified in 2000 and 2002, dates had been published from Java that hinted at the very early (1.8- to 1.6-million-year) presence of *Homo erectus* in eastern Asia; together, these dates placed beyond dispute that the hominid exodus from Africa had begun almost immediately following the appearance there of the new hominid body form—vastly earlier than anyone had previously suspected.

By 2004 four hominid crania and three mandibles, including one very large one, had been recovered from the Dmanisi deposits. Now bracketed between 1.85 and 1.76 million years old, they make up a varied assortment—particularly if you include the big mandible, of which the counterpart cranium has reportedly since been discovered—but the fact that they were found pretty close together in the sediments has been taken by most observers to enhance the probability that all represent the same species. The large mandible received the name *Homo georgicus* in 2002, but the Dmanisi team has since returned to the notion that all the fossils belong to *Homo erectus,* effectively postponing any definitive decision on their taxonomic assignment.

Heterogeneous as they are, the Dmanisi fossils are actually quite distinctive; and one thing they all have in common is small brains. Cranial volumes vary between 600 cc and 775 cc (they mostly cluster around the low end of this range), making all of them significantly smaller-brained than the Turkana Boy. And they are smaller-bodied, too: two partial skeletons, including that of an adolescent associated with one of the skulls, indicate that these hominids were diminutive compared to their Kenyan relative—so, relatively speaking, their brains may not have been quite so small. Estimates of stature based on individual long bone lengths came in at a rather short four feet ten inches to five feet two inches, but the shape of the bones themselves is said to be strongly suggestive of modern body form, with major elements showing much greater anatomical resemblance to the Boy than to any australopith.

The stone tools found alongside the Dmanisi hominid fossils resembled those used by contemporary *Homo ergaster* in east Africa: simple knobbly rock cores with sharp flakes knocked off them, barely distinguishable from the very first stone tools ever made. This was evidently a useful, adaptable, and durable technology—primitive tools like these continued to be made for a million years into the future—and the Dmanisi finds confirmed that it was not an improved toolkit that had permitted hominids to move out of Africa for the first time. Nor, if one may take the evidence of brain volumes at face value, were the hominids who made the tools any smarter than their African progenitors. So evidently, with the elimination of technology and expanded brains as potential enabling factors, it was the radical shift in hominid body form that had made the difference. Environmental change may also have been involved, the general drying-out of the climate spurring an expansion of the kind of habitat that was congenial to the new hominids: the dispersal of several other mammal species into Eurasia from Africa at around the same time indicates that environmental changes were afoot in southwestern Asia, as well as in the parent continent. Nonetheless, it's evident that the new hominids were significantly adaptable. For example, to judge from the kinds of mammals that were living there the Levant, the area at the eastern end of the Mediterranean that connects northeastern Africa to Eurasia via the Sinai Peninsula, was covered at this time primarily by Mediterranean woodlands. This environment was sufficiently different from that offered by tropical Africa to suggest that the emerging hominids must have been able to cope with a wide environmental range.

At Dmanisi itself, pollen studies indicate that just prior to its first occupation by hominids, southern Georgia had enjoyed a warm, damp climate supporting a rich mixed habitat of forested and grassy areas. But by the time the hominids actually appeared there a cooling and drying trend had begun to set in, expanding the areas of grassland and transforming the humid forests into formations of much more arid and crackly aspect. This is an environment that would have presented the hominids with a lower abundance of plant foods than their ancestors had benefited from back in Africa; but there is also evidence at Dmanisi of an extensive mammal fauna, including large herds of herbivores that the hominids must have exploited in some way—as bashed-in and cut-marked mammal bones confirm.

One new aspect of life that would have confronted the hominids arriving at Dmanisi is a pronounced seasonality in temperatures as well as in humidity; and this climatic variation would have deeply affected the resources available to them at different times of year. This was not an easy environmental transition to make, and it almost certainly could not have been managed by a typical primate species. This enhances our picture of the Dmanisi hominids as rugged, adaptable generalists, able to cope with rapidly fluctuating conditions. Evidently the key to early hominids' success in Eurasia, then as now, was the unusual flexibility of behavior that had also been the hallmark of their African ancestors.

A particular surprise offered by Dmanisi came with the discovery of the fourth hominid skull to be recovered there. Known as D3444, this skull had belonged to an aged individual, thought to have been male, who was missing all but one of his teeth. It is not unusual to find fossil skulls that have lost their teeth subsequent to the death of their owner; but in the case of D3444 most of the missing dental elements had vanished long before his demise. Nearly all of his empty tooth sockets had already shriveled away, in a process that would have taken several years.

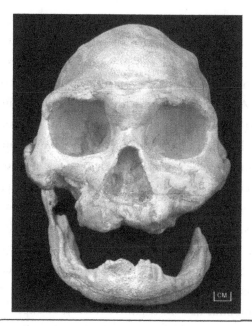

Front view of the toothless hominid skull from Dmanisi (D3444/D3900), about 1.8 million years old. This aged individual, presumed to have been male, had lost all of his teeth but one well before he died. It is thought that his survival may have required extensive help from other individuals, suggesting a social milieu of considerable complexity. Photograph of cast by Jennifer Steffey.

Especially if he and his kin had subsisted on a largely meat-based diet, this old male would have had great difficulty chewing his food; and the Dmanisi team believes that he would probably have starved without extensive help from other members of his social group (although he might conceivably have pounded meat with one of those rock cores to make it softer). Still, the argument that this highly disadvantaged individual had benefited from the long-term compassion of his relatives is, all in all, a plausible one: occasionally a chimpanzee manages to survive for a long period without teeth, but chimps eat a much softer diet than is probable for the Dmanisi hominids.

D3444 is by far the earliest example we have of an ancient hominid who somehow contrived to survive over a lengthy period with a major handicap. Indeed, the next oldest examples of disadvantaged individuals (these ones with evidence of cranial and brain deformities) are as much

as a million years younger. That the aged Dmanisi male was capable
of at least partly compensating for his physical disability by presumably cultural means has vague but powerful implications for cognitive
complexity. What's more, if the Dmanisi researchers are correct in their
surmise, D3444 also furnishes us with the first putative instance of social concern in the hominid record. Evidence for human empathy of
this kind only becomes abundant very much later in time; but, given
the spotty nature of the earlier record, this is maybe hardly surprising.
What's more, compassionate behaviors are clearly as deeply ingrained
in the human psyche as their opposites are; and it is even possible that
we may glimpse the deep roots of such expressions in the consolation
chimpanzees often offer to wounded or oppressed groupmates. What
the apes most conspicuously lack, however, is the technical capacity to
implement assistance; and it seems entirely reasonable to conclude that
the Dmanisi hominids had the cognitive reserves to express their fellow-
feeling in the form of material support. At the point when hominids first
entered Eurasia they were evidently already beings of empathy, as well
as of considerable resource and complexity.

MEANWHILE, BACK AT THE RANCH . . .

While hominids were busy spreading into other regions of the Old
World, taking long-established ways of doing business with them, those
that stayed behind in the parent continent were not standing still, at
least technologically. As in Eurasia, the old ways continued—technologies have always overlapped in time, as they continue to do today—but
at about 1.5 million years ago archaeologists in Africa (and recently in
India, too) begin to pick up regular evidence of an entirely new concept
in stone tool making. For a million years, and probably more, the key
idea in making stone implements had been to produce a smallish flake
with a usable sharp edge; and it hadn't really mattered what those flakes,
or the cores they were chipped from, had looked like. There was no
aesthetic behind them; no notion even of basic form. The tool-making
concept was totally a functional one: get that cutting edge.

Quite soon after *Homo ergaster* had come on the scene, however,
all this changed dramatically with the appearance of what is known as

"handaxes." These stone tools are the emblem of what is known as the "Acheulean" culture, so-named for the site in France at which it was first recognized. Those handaxes are pretty late, however—they are dated to well under half a million years ago—and the earliest examples of this tool type are currently known from a site in Kenya dated to some 1.78 million years ago. At that remove in time, however, handaxes are rather crudely made and extremely rare. They did not become a routine feature at archaeological sites until several hundred thousand years later.

Handaxes are much larger tools than their predecessors, and involve an entirely new notion of what a tool was. To make a handaxe, a stone "core" (in later times, itself a large flake) was elaborately shaped by multiple blows to both sides into a flat, symmetrical teardrop form. They are typically about eight or nine inches long, but may occasionally be much larger. Sometimes these tools are quite pointed, in which case they are called "picks"; in other instances they may be truncated by a straight edge, and are known as "cleavers." But the basic teardrop handaxe shape is quite uniform, and huge quantities of these implements were produced throughout the African continent and eventually beyond.

A "cleaver" (left) and a "handaxe" from St. Acheul, in France, from which the Acheulean industry was named. Photo by Willard Whitson.

The handaxe proved to be an extremely durable form: it was produced with little conceptual change, although with some later refinements in manufacturing, for well over a million years. Indeed, this implement has earned the sobriquet of "Swiss Army Knife of the Paleolithic" for its evidently many uses. Studies of how handaxes became worn in the course of such uses have shown that they were employed for tasks as diverse as cutting tree branches, slicing meat, and scraping hides. And the stability of the handaxe's form indicates just how all-purpose these tools were, even as the habitats their makers occupied changed dramatically from moist to arid and back again, sometimes on incredibly short time scales.

Making Oldowan flakes had required a considerable sophistication in choosing the stone to be chipped: coarser-grained rock was less suitable for producing or holding that slicing edge. Volcanic glass, flints, and cherts, or even fine-grained lavas would do the trick, and these materials were assiduously chosen when possible. The earliest stone tool makers clearly knew good rock when they saw it, and as we've already seen they often carried it around for long distances in anticipation of needing it in places where none might be available. But the handaxe makers faced a yet more complicated situation than the Oldowans had. Not only did the rock have to be of the right *kind,* but the individual piece of rock had itself to be suitable, free of flaws that would foil the predictable flaking sequence needed to produce a shaped utensil. So not only did the toolmaker need to be able to "see" the finished form in the rock before flaking started, but he or she needed to ensure that the core itself was sufficiently homogeneous to support the complex sequence of actions necessary. Clearly, this has profound if unclear implications for Acheulean mental abilities.

I've already mentioned that one of the most insuperable problems we have in our efforts to understand the cognitive background of any radical innovation such as handaxe making, is that we modern humans find it virtually impossible to imagine any state of consciousness other than our own. Even with the greatest effort of will we simply cannot put ourselves in the cognitive place of our predecessors, because the cognitive systems of those early hominids were clearly not simply scaled-down versions of ours. So we can't get to where the early handaxe makers were by mentally ratcheting down our IQs a notch or three: and we will certainly be on the wrong track if we think that Acheulean tool makers were simply like us, but (because they had smaller brains) dumber. Indeed, if that had been the case they would almost certainly have had a very difficult time getting by. Doing business the way we do it demands the specific kind of smarts that we have; and our symbolic way of processing environmental stimuli seems to have been a remarkably recent acquisition. Without any doubt, the early handaxe makers' subjective experience of the world, and their way of dealing with information coming in from it, were *different* in some major qualitative way from ours.

Limited as our speculations have to be, however, the conclusion is inescapable that the invention of the handaxe must have represented— or at least have reflected—a cognitive leap of some kind relative to the bipedal apes that had made the first stone tools. Making uniform objects according to a set of rules, as the handaxe makers did, implies obeisance to a collective appreciation of what is good and appropriate, and it has sometimes been considered to mark the boundary between "proto-human" and "human" behaviors. But just what did the cognitive change implied here mean in terms of what was actually going on in the heads of the hominids concerned? What did it reflect in terms of the way they apprehended and responded to the world? Unfortunately, there is nothing in the material record to suggest any answers.

Uncertainty as to what was going on is exacerbated by several other factors. For one thing, the invention of the handaxe seems to have taken place after *Homo ergaster* appeared. This is actually not surprising because, as I remarked in chapter 6, the technological advances that provide our best clues to cognition at this stage must have been made *within* a hominid species, if for no better reason than there was nowhere else for them to occur. Clearly, the intellectual potential for envisioning that a specific, realizable teardrop form lay within a lump of stone must have been present in the physical brains of the handaxe makers *before* they started expressing it. Still, the identity of the inventors of this new technology remains a bit hazy, if only because of the probability that there is more than one distinct species among the spectrum of hominid fossils generally allocated to *Homo ergaster* (let alone to the all-embracing *Homo erectus*); and, if so, we have no idea of who was doing what. The current state of our knowledge merely allows us to be certain that a spirit of innovation was astir among early *Homo* in the African continent beginning well over 1.7 million years ago. And this is quite probably the most important thing to know, especially since there is no evidence that more than one hominid *grade* was involved in the process of technological advancement.

Whatever the details of the transition, exactly how the fulfillment of the neural potential underwriting the invention of the handaxe affected other aspects of the Acheuleans' lives remains anyone's guess. For although a lot of handaxe sites are known, the kinds of activities performed at them seem, at least early on, not to have been hugely different in material terms from those documented at older sites. There is one major exception to this, though. Previous toolmakers had typically made their stone artifacts at butchery places, as and when they were needed. Tools at such places are not usually found in great abundance because of the small quantities of stone that the makers could reasonably carry around for flaking on the spot. In contrast, handaxes were often made in huge quantities, at "workshop" localities, often in close proximity to good sources of appropriate rock. Perhaps the most famous such place is at Olorgesailie, in Kenya, where literally thousands of million-year-old stone tools were found littered in a small area of the ancient landscape. This concentration of tools implies an entirely different approach to life than had been taken by the Oldowan tool makers, including the earliest *Homo ergaster*. It even strongly suggests that some degree of specialization in social and economic roles existed among members of the group.

The suggestion has also been made, though it is quite controversial, that the sites containing unusual quantities of handaxes hint at ritual gatherings, and at a social rather than purely utilitarian function for at least some of these tools. This remains pure speculation; but it is possible to draw this conclusion with more confidence from the spectacularly large size of some of the handaxes found at places like Isimila, in Tanzania. These tools were far too big and heavy to have been used for routine chores, and they have invited speculation that their uses were instead ceremonial. And while this inference may be a little too loaded with implications of our own style of humanity, it is quite possible that such tools were made in a playful spirit, or perhaps even a competitive one of showing off at large social gatherings. Expressions like these make it all the more frustrating that we have so little other supporting evidence about Acheulean lifestyles—which may well have become more complex with the passage of time: Isimila is quite late.

At Olorgesailie we also have the best current physical candidate for an actual maker of early handaxes. Not far away from the rich tool site, at the same stratigraphic level, were found fossil bits of a very small individual—far more diminutive than the Turkana Boy—who the excavators of the site suspect may have been a member of the population to which the toolmakers belonged. They described the specimen as a *Homo erectus,* but in truth this is much more because of its date than its morphology. The cranial fragments look nothing like the type specimen from Java, or, for that matter, the Boy. Still, under current standards it is entirely reasonable to attribute it to an early member of the genus *Homo;* and while a guesstimate places its brain size at under 800 cc, this lowish figure is nonetheless within the *Homo erectus/Homo ergaster* range, especially considering the individual's petite size.

BRAINS AND BRAIN SIZES

With *Homo ergaster* and *Homo erectus* we confront for the first time an episode of marked brain-size enlargement among the hominids, and this issue is a uniquely important one to confront. The brain sizes of fossil hominids have attracted a huge amount of attention, not simply because our large brains have for long been our most vaunted organs (at least as they distinguish us from other animals), but also because brain volume is easy to quantify as long as you have a fossil cranial vault that is sufficiently well-preserved to measure, or from which to make an estimate. One fascinating thing about hominid brain sizes over the last couple of million years is that, without any doubt, they show a striking trend toward enlargement with time. Brain sizes had more or less flatlined during the long tenure of the australopiths. The very latest australopiths seem on balance to have had slightly larger brains than the earliest ones; but even if this doesn't simply reflect body size, the difference is too small to be worth noting. But once the genus *Homo* was on the scene, everything changed. On average, the later in time a fossil member of the

genus *Homo* is, the larger its brain is likely to be. This really is important, because the way we process information in our heads most clearly
demarcates us modern humans from all other creatures on the planet;
and our cognitive abilities are certainly dependent on our large brains,
even if size is not in itself the whole story.

So brain size is without doubt a critical factor in human evolution.
But we have to be careful how we interpret it. Particularly under the
gradualist dictates of the Evolutionary Synthesis, paleoanthropologists
have often been tempted to simply join the dots into a single continuum
of brain size increase. Two million years ago, our ancestors' brains were
basically ape-sized; a million years later, they had doubled in volume;
and today they are twice as large again. What could be more suggestive
of an inexorable trend, as smarter individuals out-reproduced dumber
ones? And what, looking back, could be a greater compliment to our
current finely burnished species? When you think about it, this really is
the ultimate paleoanthropological feel-good formula.

But there are other ways to look at the brain size picture. For a start,
even though we don't have anything like the number of fossil hominid
braincases we'd like, we do have enough to know that at any one point
in time brain sizes varied widely. Among the australopiths, we're in the
fairly tight range of about 400 cc to 550 cc. Among the earliest species
of *Homo*, in the period following about two million years ago we're
looking at a range of some 600 cc to 850 cc; and by around half a million years, give or take, the range has broadened to around 725 cc to
1,200 cc.

We are also looking at a pretty impressive if often unacknowledged
variety of morphologies. Take, for example, four eastern African crania,
all dated to roughly one million years ago: that tiny individual from
Olorgesailie (reckoned to have had a brain capacity of less than 800 cc);
a cranium from Buia in Ethiopia (750–800 cc); a skullcap from Daka in
Ethiopia (995 cc); and a braincase from Bed II at Olduvai Gorge (1,067
cc). All have been allocated to the species *Homo erectus*; but all of them
look decidedly different, not only from the Javan type specimen of the
species, but from each other. There is clearly more going on here than a
simple lump categorization as *Homo erectus* or even *Homo ergaster* can
reflect, for none of them looks much like the Turkana Boy, either.

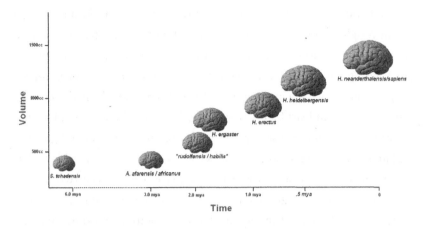

A crude plot of average hominid brain sizes against time. After an initial flatlining, this plot appears to indicate a consistent enlargement of the hominid brain over the last two million years. But it is important to bear in mind that these brain volumes are averaged across an uncertain number of different lineages within the genus Homo, *and that it is likely that what the plot actually reflects is the preferential success over this period of larger-brained hominid species, rather than steady increase within a single lineage. Illustration by Gisselle Garcia.*

A perfect example of the conflicted mindset that results from the tension between evidence and tradition came a few years ago, when a team working in the Turkana Basin described two new hominid fossils from the Ileret area to the east of the lake. One was a lightly built braincase (691 cc) some 1.55 million years old, that they allocated to *Homo erectus* even though it bears none of the morphological hallmarks of the Javan type specimen. The other was a piece of upper jaw, perhaps a hundred thousand years younger, that they assigned to *Homo habilis*. The researchers hailed these finds as evidence that at least two distinct lineages of the genus *Homo* had occupied the Turkana Basin at about the same time, emphasizing diversity among the hominids of the period. Yet their allocation of the braincase to *Homo erectus* could only have borne any conceivable logic in the context of the view that *Homo erectus* is the middle grade of a single, worldwide, variable, and gradually evolving hominid lineage—precisely the construct that they ostensibly wished to undermine.

Of course, with modern humans presenting us with brains ranging from roughly 1,000 cc to 2,000 cc in volume, we can hardly use past brain size variation alone to reject the notion of a single variable human lineage that consistently gained in brain volume with time. But the huge variation we see in the morphology of the skulls those brains were contained in is at least highly suggestive. And if multiple hominid species *were* out there in the past—species whose brain size ranges and geological lifespans are, regrettably, unknown to us—then it is just as likely that the trend we see toward increasing brain size over the last two million years is due to the greater competitive success in the ecological arena of larger-brained hominid species, as that it is due to the reproductive success of larger-brained and thus smarter individuals.

A scenario involving the consistent triumph of larger-brained species might be taken to suggest that the pressures favoring hominid brain expansion over time were essentially ecological, and thus external to the species themselves. Nonetheless, there is one important observation that suggests that members of the genus *Homo* have been consistently predisposed in some way toward brain size increase: brain enlargement has occurred independently in at least three lineages within the genus. The earliest *Homo erectus* in Java, dating from perhaps more than 1.5 million to a bit under one million years ago, have brains ranging from around 800 cc to a bit over 1,000 cc in volume. A later Javan group, poorly dated but maybe about a quarter of a million years old, comes in at 917 cc to 1035 cc; and the latest Javan *Homo erectus* group of all, perhaps no more than 40 thousand years old, varies from 1,013 cc to 1,251 cc. Similarly, *Homo sapiens* and *Homo neanderthalensis* diverged from a smaller-brained common ancestor well over half a million years ago, and independently gained their comparably sized large brains. Thus, a 600-thousand-year-old collection of Spanish fossils foreshadowing the Neanderthals had brains some 1,125 cc to 1,390 cc in volume, against the later Neanderthal average of 1,487 cc.

Given that *Homo erectus* lived in tropical eastern Asia, the Neanderthals in Ice Age Europe, and the precursors of *Homo sapiens* in Africa, it is hard to see a common environmental thread in the trend toward bigger brains that all three lineages independently followed. Somehow, very early on in its evolution, the genus *Homo* must have acquired some underlying predisposition, biological or cultural, toward brain enlargement. Identifying that factor will be essential if we are ever to have a full account of how we became the extraordinary cognitive entity we are—even though, as we will see, while a large brain is clearly a necessary condition for our unique modern cognitive style, it is not a sufficient one.

Still, predispositions notwithstanding, there is nothing inevitable about the enlarging brain in *Homo*. We were all forcibly reminded of this by the recent discovery of the remarkable "Hobbit" at Liang Bua Cave on the Indonesian island of Flores. The best specimen of this extraordinary hominid, technically known as *Homo floresiensis,* is the skeleton of a tiny individual, dubbed LB1. In life LB1 had stood not much more than three feet tall and, though bipedal, had possessed rather unusual body proportions. LB1 also has a head with a tiny vault that had contained a brain of maybe as little as 380 cc in volume—fractionally smaller even than Lucy's, the smallest australopith brain known. What's more, perhaps most bizarrely of all, the individual had lived only about 18 thousand years ago.

Predictably, the announcement of the entirely unexpected Hobbit was greeted by enormous controversy. The scientists who described LB1 thought it might be a dwarfed descendant of a population of *Homo erectus* that had somehow contrived to reach Flores in the remote past. This is not in itself implausible: "island dwarfing" of mammals and reptiles is not unusual on smallish isolated landmasses such as Flores—and indeed, bones of a tiny elephant are found in the same cave deposits that yielded LB1. But there is little about its anatomy to suggest any close affinity to *Homo erectus,* and LB1's brain is much smaller than you would expect to find as a result of normal processes of dwarfing, even from the typically modest-brained *Homo erectus.* Several different authorities have suggested, alternatively, that LB1 is simply the skeleton of a pathological modern human; but none of the suggested disease conditions fits the case well enough. As more is learned about this specimen, the more likely it appears that LB1 and its kind will in the end prove to be descendants of extremely early émigrés from Africa, preserving archaic features that may well eventually help us learn just what those émigrés were like.

Meanwhile, independent of whether or not some degree of island dwarfism is indeed involved in this particular case, LB1 tells us that, larger patterns apart, time and brain enlargement need not necessarily be synonymous among members of the genus *Homo*—if, that is, it is appropriately assigned to our genus at all.

EIGHT

THE FIRST COSMOPOLITAN HOMINID

The systematic picture among fossil hominids of the period around a million years ago remains rather unclear, because relevant fossils in the African center of innovation are few and far between and widely scattered. But it clarifies considerably about 600 thousand years ago, when we get the first indications of what was the world's first Old World–wide hominid species, *Homo heidelbergensis*. This species is based on a mandible that was discovered back in 1908 in a gravel pit at Mauer, near the German city of Heidelberg, but that was only recently dated to 609 thousand years ago. On its own the Mauer jaw was a bit puzzling, but fortunately it matches well with jaws present in a fossil sample from Arago, a cave in the French Pyrenees that also preserves a face and associated vault bones dating to around 400 thousand years ago. Thus armed with much of a skull, we can be confident in also assigning to *Homo heidelbergensis,* or to something very much like it, a 600-thousand-year-old partial cranium from Bodo, in Ethiopia, and crania from Kabwe in Zambia, Petralona in Greece, and Dali and Jinniushan in China, plus other less complete specimens from Africa and elsewhere. Dating is rather poor for most of these fossils, but all seem to fall in a broad range between about 500 thousand and a bit less than 200 thousand years ago. Equally unfortunately, we can't say much about the body structure of *Homo heidelbergensis,* since bones of the body skeleton are few and far between (except in the Chinese Jinniushan specimen, details of which are not yet fully available to the scientific community). Still, what we do know suggests a build along the basic modern body plan, if very robust and differing from us in various details that foreshadow those of the Neanderthal skeleton, which we'll learn more about shortly.

Given what we currently know, it seems most likely that *Homo heidelbergensis* arose in Africa and then spread out of that continent, just as the first hominid émigrés had done before it. The details of its origin continue to elude us among the welter of hominid morphologies so tantalizingly suggested in the post-one-million-year record; but what is uncontestable is that, with the arrival of *Homo heidelbergensis,* we are entering new adaptive territory. In its cranial morphology this new species once more anticipates the future at least as much as it echoes the past, with robust but flatter faces with shorter tooth-rows than its predecessors have, tall eminences above its eyes, and capacious cranial vaults ranging in volume from 1,166 cc to 1,325 cc. Here we

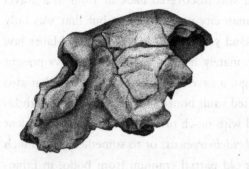
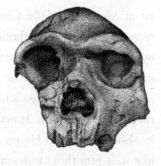

The partial cranium of Homo heidelbergensis *from Bodo in Ethiopia. At 600 thousand years old this is one of the oldest known representatives of its species. Drawing by Don McGranaghan.*

are comfortably within the modern brain size range, if still somewhat below today's average.

All of the brain endocasts of this species that have so far been described are said to show expanded Broca's areas; but beyond this the paleoneurologists who study them have been disappointingly mute, although they have generally been more impressed by the endocasts' similarities to modern brains than by the differences they see. Still, while the endocasts are said to show cerebral asymmetries of the kind one finds between the left and right halves of the brain in modern people, the prefrontal areas of the cortex that lie directly above our eyes (though in *Homo heidelbergensis* they lie variably above and behind them) are usually characterized as broad and flat compared to those of *Homo sapiens*—exactly as you'd expect from looking at the skull that enclosed them.

In modern humans the prefrontal cortex is vital to our complex cognition, governing such key areas of mental activity as decision making, the manifestation of social behaviors, and the expression of personality traits; and it seems reasonable to conclude that its role was broadly similar in *Homo heidelbergensis*. But exactly what the difference in external appearance between the prefrontal regions of the two species means in terms of their exact function is not at all clear, especially since we have no idea how this brain area was internally organized and connected to adjacent structures in *Homo heidelbergensis*. Equally, we really can't say just what kind of mental edge the marginally greater brain size of *Homo sapiens* might by itself give us over the cognitive condition of *Homo heidelbergensis*. All of which means that, though we can legitimately conclude on the basis of brain size alone that *Homo heidelbergensis* was somehow "smarter" than its predecessors, there is no way to determine the details of this increase in intelligence by any form of direct observation. Once more, we have to turn to the indirect proxy indicators furnished us by the archaeological record.

The oldest fossil specimens we have of *Homo heidelbergensis* either lack any archaeological context (Mauer) or have pretty archaic associations (Bodo). Indeed, though stone tools are quite common in the Ethiopian deposits from which the Bodo skull was recovered, they are of Oldowan type, and handaxes are conspicuously absent—a million years after their invention, and despite their occurrence in underlying (older) layers. Once again, we see no correlation between the appearance of a new kind of hominid and technological innovation. Interestingly, though, the Bodo specimen bears cut-marks made when the bone was fresh, as if the skull had been deliberately defleshed. What this means is uncertain, beyond the clear inference of hominid intervention. Cannibalism is unlikely, because the marks appear on the face and forehead, areas that would not have yielded much that was edible; but equally we should be careful not to read into these marks evidence of ritual in the sense in which it is familiar to us today.

Fortunately, a number of archaeological sites in Europe fall within the timespan of *Homo heidelbergensis,* and help us flesh out the story of what hominids were up to at the time. One particularly interesting hominid occupation locality is at Terra Amata, in a suburb of Nice on the French Mediterranean coast. Some 380 thousand years ago the terrace on which Terra Amata lies was an ancient beach, to which a small group of hunters repeatedly returned (alas, without leaving any evidence of themselves). There they built a number of large shelters, as indicated by oval rings of large stones that had served to anchor rows of saplings that were embedded in the ground and brought together at the top. Whether or not these shelters were converted into true huts by covering them with hides is unknown, but seems likely. A break in the ring of stones defining the best-preserved of these structures indicates not only its entrance, but also a place through which smoke would have escaped from a fire built directly inside. This fire burned in a shallow scooped-out hearth, in which archaeologists found both blackened cobbles and the bones of animals whose meat had presumably been cooked. This hearth at Terra Amata is perhaps the oldest firm evidence we have for the domestication of fire following that extraordinary 800-thousand-year-old outlier in Israel, and it announces the advent of fire as a regular part of the documented human behavioral repertoire. For, from Terra Amata times on, hearths become an increasingly common feature of hominid occupation sites.

Building structures and routinely using fire represent significant steps in the direction of modern behavioral patterns; yet at the same time, the numerous stone tools found at Terra Amata are remarkably unsophis-

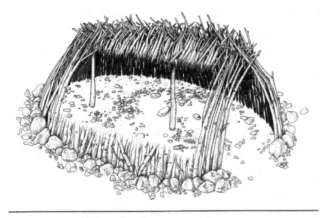

Reconstruction of the earliest documented shelter, some 350 thousand years old, from Terra Amata in France. A shallow hearth has been scooped out just within the entrance indicated by a break in the ring of reinforcing stones. Concept by Henry de Lumley; drawing by Diana Salles.

ticated. There are no bifacially shaped handaxes, and most of the tools are simply produced flakes. The crudeness of the tools may be accounted for by the fact that the local rock (cobbles of silicified limestone) did not provide good raw material for cutting tools, though it may be significant that the Terra Amata hominids had carried in some pieces of red and yellow pigment from distant sources, evidently attracted by their aesthetic qualities.

An equally unusual and interesting site is the locality of Schoeningen, lying in an anaerobic peat bog in northern Germany. This unusual environment has preserved the earliest evidence we have of wooden tools, as well as the remains of ten butchered horses and numerous other mammals—though sadly (for us), not those of the hominids who had hunted them. Dead wood is a fragile material that rarely lasts beyond a few decades, or centuries at best; and although it must have been used as a resource by hominids back into the deep recesses of time (australopiths very likely used digging sticks), the preservation of several long and carefully-crafted spears some 400 thousand years old is little short of miraculous. Indeed, before this find the earliest wooden implement was a yew spear tip, a mere 125 thousand years old, that had been found at the German site of Lehringen amid the ribs of a straight-tusked elephant that it had presumably been used to hunt.

In a cool-temperate setting such as that at Schoeningen when the deposits there were formed, there is a fairly restricted variety of naturally occurring plant resources suitable for hominids to exploit. As a result, one might have suspected that a substantial meat component in the diet was necessary to sustain hominids at this latitude. But the sophistication of the spears nevertheless came as a surprise to archaeologists, many of whom believed in the mid-1990s, when the find was made, that any spears used by hominids that far back would have been of the handheld thrusting kind. Yet the Schoeningen spears clearly resembled throwing implements, six-plus feet long and shaped like Olympic javelins, with their weight concentrated toward the front. Made from spruce, their tips were finely shaped to a sharp point; and although it has been objected that a wooden point would likely have bounced off the thick hide of a large mammal unless the projectile had been thrown from very close range (which would have more or less eliminated its advantage relative to a manual thrusting spear), it is claimed that these spears were nonetheless designed to be thrown, implying the advent of sophisticated ambush hunting. And whatever the hominids' exact technique, the ample remains of a range of butchered large mammals testifies to the hunting efficacy of the Schoeningen hominids.

The Schoeningen site yielded another "first" as well, and probably a more significant one. A major innovation in the technological history of hominids was the invention of the "composite" tool, made from more than one component. This can lead to a huge increase in efficiency, as anyone who has ever tried using a hammer without a handle will know. At Schoeningen, in addition to a variety of flint flakes, three worked sections of fir branch were found, several inches to a foot long, each of them notched at one end. It is believed that these branch sections had formed the handles of tools that were tipped with flint flakes glued or bound into the notches. There are no traces of potential glue on any of the Schoeningen stone tools; but we know from a bit later in time that the Neanderthals, for instance, used natural resins in this role.

If the Schoeningen wood segments were indeed the handles of composite tools, then this improvement preceded the next major advance in stoneworking itself, at least in Europe, by 100 thousand years. Sometime between about 300 and 200 thousand years ago, "prepared-core" implements were introduced, representing a radically new approach to the shaping of stone: one that required a high quality, predictably fracturing stone such as flint or chert. The toolmaker elaborately shaped a stone core on both sides, using numerous carefully positioned strikes and usually employing a "soft" hammer of bone or antler, until a final blow would detach a finished tool. This would have a more or less continuous cutting edge nearly all the way around it. The core could be discarded, or another flake could be knocked from it; and the flake(s) could then, if desired, be modified into a specific shape, perhaps for scraping or cutting.

In this new means of tool production we clearly see a new layer of cognitive complexity introduced into the hominid behavioral repertoire. Not only did the toolmaker have to envision the form of the finished tool before the process began, but he or she had to be able to plan and conceptualize several stages ahead, instead of heading straight for the desired shape. Whether this new approach to making tools was invented in Europe or Africa or independently in both regions is unclear, as is the exact identity of the inventor(s); but it represented an idea whose time had come, and it was an innovation that occurred on *Homo heidelbergensis*'s watch. Interestingly, in the parent continent of Africa there is early evidence of the production, using stone hammers, of "blades"—basically, flakes with parallel sides and more than twice as long as wide—at a site in Kenya dating to over half a million years ago, also within the time span of *Homo heidelbergensis*. The appearance of blades in Africa that long ago is particularly intriguing because such implements, struck from cylindrical cores, are only found in Europe many hundreds of thousands of years later, with the arrival of fully cognitively modern humans. Blade production is no mean feat, involving as it does a complex sequence of actions, together with a firm grasp of the properties of the material being worked; and, whatever exact species the Kenyan early blademaker belonged to, he or she had completed a very demanding cognitive task.

The tenure of *Homo heidelbergensis* on Earth, approximately between 600 and 200 thousand years ago, thus witnessed a large number of lifestyle and technological innovations among hominids. And though we cannot identify the authors of these innovations with any certainty, we can with reasonable confidence attribute them to *Homo heidelbergensis* or something very like it. These were hardy, resourceful folk, who occupied and exploited a huge range of habitats throughout the Old World through the deployment of an amazing technological and cultural ingenuity. They were adroit hunters who pursued large game using sophisticated techniques, built shelters, controlled fire, understood the environments they inhabited with unprecedented subtlety, and produced admirable stone tools that at least occasionally they mounted into composite implements. Altogether, they lived more complex lives than any hominids had ever done before them.

Yet in isolation we cannot confidently read symbolic thought processes into any techniques of stone-knapping; and throughout the period of *Homo heidelbergensis*'s tenure no hominid produced anything, anywhere, that we can be sure was a symbolic object. Perhaps only a couple of very late items even qualify for consideration in this category. One of these is a "Venus" recovered at Berekhat Ram, a 230-thousand-year-old site on the Golan Heights excavated by Israeli archaeologists in 1981. The "Venus" is a small pebble that is vaguely shaped like a human female torso. It has been argued that this object's anthropomorphic aspects have been amplified by three deliberately incised grooves, though it remains uncertain that any purposeful human action was involved. The second contender is a couple of small round perforated disks of ostrich eggshell from Kenya that may arguably have been objects of personal adornment (and therefore symbolic), and are even more arguably up to 280 thousand years old. But both the dating and the interpretation are speculative, and there is certainly nothing in the material record to suggest that the symbolic manipulation of information was in any way a regular part of the cognitive repertoire of *Homo heidelbergensis*. Had it been, we would surely expect to see more material evidence of it.

Homo heidelbergensis was certainly remarkable, and in their day its members were undoubtedly the most intelligent creatures that had ever existed on Earth. But although we can see numerous similarities to ourselves in them—as indeed it's also easy to do, albeit to a lesser extent, in chimpanzees—members of *Homo heidelbergensis* were not merely simpler versions of us. If I had to wager a guess, it would be that the intelligence of these hominids, formidable as it may have been, was purely intuitive and non-declarative. They neither thought symbolically as we do, nor did they have language. As a result, we can't usefully think of them as a version of ourselves, certainly cognitively speaking. Instead, we need to understand them on their own unique terms. As I've already emphasized, this is not easy to do even at the best of times; and in the case of *Homo heidelbergensis,* where the clues we have about these hominids' lives are hugely tantalizing yet so few, it is particularly difficult.

NINE

ICE AGES AND EARLY EUROPEANS

The continent of Africa has consistently been the fount of innovation in hominid evolution. But—and only partly because it has been scoured more intensively than any other part of the globe for traces of the human past—Europe also has a huge amount to offer us in terms of understanding just how it is that we differ even from our closest extinct relatives. The key to this understanding lies in the breadth of our knowledge of the endemic European hominid species *Homo neanderthalensis*. This species is better documented than any other of our extinct hominid relatives, and, very importantly, its members boasted brains as large as our own, if not even fractionally larger. The Neanderthals are thus ideally placed to act as a sort of mirror to our own uniqueness: an alternative take on the theme of the large-brained hominid that helps us to gauge whether or not our vaunted mental prowess is simply a sort of passive byproduct of the metabolically expensive "more brain is better" theme that—for whatever reason—seems to have dominated the history of the genus *Homo*. To make the comparison more complete, we can contrast our own behaviors with what we can infer of theirs in unaccustomed detail, because the Neanderthals left us an unusually complete material record of their existence. By making this comparison we may hope to gain some perspective on exactly what it is about us that has made us the lone hominid in the world today, and a species that interacts with that world in an unprecedented way. But before we look at what the Neanderthals add to the human story, let's quickly look at the climatic backdrop against which the events of later hominid evolution took place. For environmental change (on various scales) has been the most important single driver in the evolution of the organic world, human beings not excepted.

THE ICE AGES

We've already seen that, well before the time at which our genus *Homo* originated, climates worldwide were undergoing the gradual deterioration that spurred the development of the relatively open African habitats colonized by the early hominids. This trend received a huge impetus about three million years ago, when the collision of North with South America produced the Isthmus of Panama. The new land barrier blocked the circulation into the Atlantic Ocean of warm Pacific water, producing an acceleration of the cooling and drying trend in Africa, and initiating the formation of an ice cap in the Arctic. We see the results of this event dramatically expressed in the African fossil record beginning around 2.6 million years ago, with a proliferation of grassland-adapted grazing mammals and the disappearance of older browsing forms. Some authorities believe that the environmental shift reflected in faunal change around this time was the most significant stimulus to the emergence of our genus *Homo;* and whether or not this was actually the case, it is certainly true that the underlying event ushered in a new climatic cycle that exerted a profound influence on later phases of hominid evolution. In Africa temperatures remained relatively warm, but the continent was deeply affected by major fluctuations in rainfall. In Eurasia the effects were greater yet, since the more northerly latitudes into which hominids began moving some two million years ago were also influenced by significant excursions in temperature.

The initiation of the Arctic ice cap at around 2.6 million years ago marked the beginning of the "Ice Ages" cycle of alternating glacial and interglacial episodes, as the ice caps at both ends of the Earth regularly expanded and contracted. These fluctuations occurred because of differences in solar radiation received at the planet's surface due to variations in its orbit around the Sun. By about a million years ago, when vast Serengeti-style savannas were becoming established in parts of Africa, this cycle had settled down into a fairly steady rhythm, swinging every hundred thousand years or so from cold troughs to warm peaks (of the kind we are experiencing today). Between the extremes, numerous shorter-term oscillations occurred. Sometimes those oscillations were very short-term indeed, rather like the "Little Ice Age"—which itself showed three distinct temperature minima—that spanned the sixteenth to nineteenth centuries.

At the peaks of cold, the Arctic ice cap expanded to cover much of Eurasia as far as 40 degrees south, and subsidiary ice caps on the Alps, Pyrenees, and other Eurasian mountain chains grew and sometimes coalesced to form formidable geographical barriers. Environments in proximity to the ice varied substantially, depending on local topographic features and how far away the ocean was. But in most places the ice masses yielded rapidly to tundra, where sedges, lichens, and grasses grew on a thin layer of soil above the permafrost and supported large populations of grazing mammals such as musk ox and reindeer. Farther south, and in sheltered areas, the vegetation grew taller, with pine forests ultimately giving way to mixed conifer and deciduous formations in which deer roamed. As the climate warmed up, the ice retreated northward and the vegetation bands followed, taking their faunas with them. In the south, broadleaf forests dominated during milder periods, giving way to Mediterranean-style scrublands in drier areas. As all this was going on geography itself changed, due to the locking up of seawater in the ice caps during colder periods. At times of maximum ice cover, world sea levels fell as much as three hundred feet compared to today's, thereby uniting such warm-period islands as Britain and Borneo with the adjacent mainland, and extending continental coastlines far seaward. In warmer times the encroaching sea doubtless repeatedly swamped many major sites of human glacial habitation.

The latest official geological determination (with which not everyone is happy) is that the start of the glacial cycle around 2.6 million years ago marks the beginning of what geologists call the Pleistocene ("most recent") epoch, which runs right up to the last major ice cap retreat about 12 thousand years ago. The time since is known to geologists as the Holocene ("entirely recent") epoch—although, human impact apart, there is no good reason for thinking we are out of the glacial cycle. However you may choose to define our genus *Homo*, it is thus a product of the Pleistocene; and the bottom line here is that our ancestors evolved in a period of increasingly unsettled environmental conditions. This was true both in the home continent of Africa, where rainfall varied dramatically on compact timescales, and in Eurasia, where vast swaths of the continent were periodically rendered uninhabitable to hominids of the time. There is thus no way in which we can realistically think of hominid evolution during the Pleistocene as a matter of steady adaptation to a specific environment, or even to an environmental trend. Instead, the story is a much more dramatic one, as tiny hominid populations were buffeted by changing conditions often retreated or became locally extinct, simply victims of being in the wrong place at the wrong time.

It's worth noting, though, that by regularly fragmenting already sparse hominid populations both in Africa and Eurasia, the Pleistocene offered ideal conditions for the local fixation of genetic novelties and for speciation. Both of these are processes that in creatures such as hominids depend on physical isolation, and small population sizes. Ice Age conditions were often tough for the hominid individuals concerned; but never had circumstances been more propitious for meaningful evolutionary change than among our highly mobile, adaptable, and resourceful Pleistocene ancestors. Taken together, this combination of internal and external factors may well account for the amazing rapidity with which hominids evolved during the Pleistocene. For there can be no doubt that the evolutionary history of hominids during this epoch was hugely more eventful than that of any comparable group of mammals of the period. We differ today far more from our earliest Pleistocene ancestors than do any other of the creatures with which we share the planet.

Ironically—because ecological generalists normally have much lower speciation and extinction rates than specialists—this rapid evolution was almost certainly due to our generalist ancestors' combination of flexibility and resilience, combined with a propensity to spread readily into new environments in a rapidly fluctuating world. The process would have been helped along by the sparse and scattered population structure that resulted from the hominids' secondary adoption of a predatory lifestyle. And very recent findings have also pointed to something quite unexpected: the possibility that, under fluctuating Pleistocene conditions, new genes may have been introduced into hominid populations by occasional intermixing between closely related and poorly differentiated hominid species.

But one important final factor that is totally unique to hominids, and which appears in some respects paradoxical, is the possession of complex culture, especially as it is expressed in technology. The exploratory inclinations of our ancestors could never have been indulged in the absence of their ability to accommodate technologically to unfamiliar and extreme conditions. Culture is usually—and justifiably—viewed as a factor that has helped insulate hominids from their environments and thus from biological selection. But in this particular context, its role in facilitating the huge geographical dispersion of thin-on-the-ground hominids may actually help explain how the genus *Homo* contrived to evolve so fast during the Pleistocene.

Early geologists constructed a chronology of the Pleistocene using physical evidence of the advances and retreats of the glaciers—such as horizontal scratches on valley sides and floors made by pieces of rock carried along in the ice, or the deposits of such rock that were dumped when the glaciers melted. But the problem was that each glacial advance scoured away much of the evidence left behind by its predecessors, and the resulting observations were a nightmare to interpret. Since the 1950s the older division of Pleistocene time into four major glacial and interglacial periods has thus given way to a chronology based on modern geochronological and geochemical analyses of long cores drilled through sea-floor sediments, or through the Greenland or Antarctic ice caps.

In both cases, the favored approach has been to measure the ratios of lighter and heavier isotopes of oxygen in the layers of the accumulating ice itself, or in the shells of microorganisms that lived in the surface waters and sank to the seafloor when they died, forming a sediment pile. This ratio provides a guide to prevailing temperatures because the lighter isotope more readily evaporates from seawater than the heavier one does. When the vapor is precipitated as rain or snow over the poles in cold times, the light isotope becomes "locked up" in glacier ice. As a result, cold oceans and the microorganisms living in them are enriched in the heavier isotope, while the lighter one is more abundant in the ice caps. And the ratio between the two in an ice or seafloor core correlates closely with the temperatures prevailing when the ice/sediments were formed. These cores provide a continuous record of fluctuations in this ratio, and thus of shifts in prevailing temperatures over time.

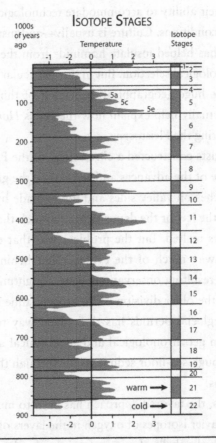

The oxygen isotope record of changing global temperatures for the past 900 thousand years, based on $^{16}O/^{18}O$ ratios in cores from the Indian and Pacific ocean seabeds. Even-numbered stages were relatively cool, odd-numbered ones relatively warm. Within each major stage there were considerable oscillations in temperature. Data from Shackleton and Hall (1989); chart by Jennifer Steffey.

From such data, paleoclimatologists have been able to identify 102 separate "Marine Isotope Stages" (MIS) since the start of the Pleistocene, and have numbered them, starting with the most recent. As a result, warm stages are given odd numbers and cold periods get even ones. We are now in warm MIS 1, the peak of the last glacial episode is represented by MIS 2, and so forth. Within each major episode of temperature fluctuation there are numerous minor peaks and troughs called "stages," some of them significant enough to have their own designations. Stage 5, for example, is subdivided into Stages 5a, b, c, d, and e, the oldest of them (5e) so warm that sea levels were several meters higher than at present.

Far back in the early Pleistocene, temperature oscillations were frequent though not very pronounced; but as we approach the present they have become more widely spaced, and more intense. Tying in particular hominid fossil sites to the marine or ice cap sequences is not always easy when absolute dates are not available; but since erratic environmental conditions usually result in frequent faunal changes, the identity of associated animal remains will often provide valuable clues. In any event, in combination with new methods of dating and independent measures of climate such as the analysis of fossil pollen and soils, we now have a pretty good notion of the tapestry of environmental challenges with which our precursors had to cope.

THE FIRST EUROPEANS

It is against an unsettled climatic and geographic background that we have to consider the early hominid occupation of Europe. Until not long ago, it was believed that early hominids first entered Europe relatively recently, certainly much more recently than the parts of southern Asia that hominid populations could reach by expanding along subtropical coastlines. The Dmanisi discoveries, right on the crux between Asia and Europe, showed that contrary to expectation the temperate zone had been penetrated very early on; and now there is direct physical evidence from an Iberian site that hominids had established themselves in western Europe by 1.2 million years ago. This evidence comes from a site known as the Sima del Elefante in the limestone Atapuerca Hills of northern

Spain, and it consists of a piece of lower jaw of the genus *Homo*, bearing a few worn teeth, that is too incomplete to be assigned to any particular species. Associated with this specimen are some mammal fossils suggesting that the hominid had lived during a relatively warm stage; and stone tools of Oldowan aspect demonstrate that it was not an advance in technology that had permitted hominids to penetrate the Iberian peninsula at this early date. As far as lifestyle is concerned, there's nothing much beyond this to go on. More as a matter of convenience than anything else, the scientists who discovered the Sima del Elefante fossil tentatively associated it with some similarly fragmentary hominid fossils from the nearby Atapuerca site of the Gran Dolina, that they had previously assigned to the new species *Homo antecessor*.

This latter species, some 780 thousand years old, is of particular interest because the Atapuerca scientists believe it may represent the common ancestor of the lineages that led to *Homo neanderthalensis* on the one hand, and to *Homo sapiens* on the other. But while the Gran Dolina does lie more or less in the right time zone for its hominids to play that role, just where they stand in the evolutionary picture remains equivocal. Indeed, it seems at least equally likely that the *Homo antecessor* fossils are evidence of an early "failed" hominid foray out of Africa and into Europe that did not have a direct ancestral connection to the Neanderthals who later established themselves in the European Peninsula. Still, if there is a direct connection back to the Sima del Elefante hominid, the first hominid occupation of Europe was a long one; and if you are looking for continuity, you might find further evidence for it in the fact that the crude stone tools from the two Atapuerca sites do not differ much.

The spot at the Gran Dolina where the hominids were found seems to represent an ancient cave entrance that was occupied by hominids during a relatively mild and humid period. And the Atapuerca scientists' claim that *Homo antecessor* represents the common ancestor of Neanderthals and modern humans is no less remarkable than their conclusion that the bone fragments assigned to this species show evidence of cannibalism. Fossil bones so far recovered from the Gran Dolina are typically badly broken, and many of them bear marks left by slicing, chopping, and scraping with stone tools, along with fractures of a kind strongly suggesting butchery. What's more, allowing for differences in anatomy among species, all of the bones, human and nonhuman alike, were treated in an identical fashion. And this implies that all of the cadavers they represent were used for the same purpose, namely consumption. There is certainly no evidence of any special or ritual treatment of the human remains. As a result, there is a strong case to be made that hominids were eating other hominids at the Gran Dolina 780 thousand years ago; and though not everyone is entirely happy with this interpretation, the argument for cannibalism is a strong one and the Atapuerca team has recently issued a robust defense of its conclusion.

Demonstrating cannibalism is just the beginning, as it raises a host of questions, foremost among them, who was eating whom, and why? To modern humans, cannibalism has all kinds of symbolic overtones, depending, for instance, on whether you are eating your kin or strangers. The Atapuerca group dismisses any symbolic implications at the Gran Dolina, emphasizing that the 11 children and adolescents represented in the butchered sample received no special handling, and that the butchery techniques involved were designed to extract the maximum amount of edible material, including the brains. Because they can find no evidence to support the idea that the butchered humans at the Gran Dolina were the victims of a single episode of group starvation—indeed, the butchery may have taken place over several tens of thousands of years, in the midst of a rich habitat—they propose that cannibalism was a regular part of the subsistence strategy of *Homo antecessor.* They even go so far as to speculate that the tender ages of those eaten might imply that they were the vulnerable victims of hunters sent out to raid neighboring groups.

Sadly, there is not much direct evidence from the Gran Dolina itself, apart from the butchered fossil carcasses and the stone implements, most of them knapped right there in the cave, that were used to dismember them. There is no indication of the use of fire, for example, or of any other activities that might have been associated with the hominid occupation—although some plant vestiges might indicate a more rounded diet than the bones alone suggest. Almost certainly, the Atapuerca researchers are right to reject any implications of ritual in the butchered hominid assemblage; and if they are correct in concluding that cannibalism of this very matter-of-fact sort was a routine component of dietary life at the time, the clear implication is that the hominids concerned did not have the kind of regard for others that is typical in modern human societies today. Wherever it has been documented in the historical record, socially sanctioned cannibalism, whether within or among groups, has always been a "special" activity, surrounded by its own rites and ambivalences. The chillingly prosaic nature of the cannibalism at the Gran Dolina implies something totally different—and to us, totally alien.

NEANDERTHAL ORIGINS

Although it is not possible to draw a straight line connecting *Homo antecessor* at the Gran Dolina with the later *Homo neanderthalensis,* yet another site in the astonishingly fecund Atapuerca region furnishes us with the best evidence we have for an early member of the Neanderthal lineage. The species *Homo neanderthalensis* itself does not show up in the European fossil record until less than 200 thousand years ago; but a stone's throw away from the Gran Dolina, the large Cueva Mayor cave has produced one of the most extraordinary phenomena in paleoanthropology, and one that gives us a marvelous insight into the early stages of the Neanderthal lineage. Well within the cave is a vertical shaft, almost fifty feet deep, at the bottom of which a small, cramped chamber has delivered the greatest concentration of hominid fossils ever discovered, anywhere. Hominid fossils are an extraordinary rarity, and paleontologists usually count themselves lucky to find just one or two. But the lead excavator of this Spanish cornucopia once remarked to me that his team was the only one in the world with the luxury of deciding how many hominid fossils it wanted to excavate in the next field season—a few dozen, a hundred—and then quit when the quota was filled. No wonder this amazing place is called the "Sima de los Huesos"—the Pit of the Bones. It's a hellishly cramped, difficult, and uncomfortable place to excavate, to be sure; but well worth every painful moment.

The Sima was initially discovered by spelunkers, who alerted paleontologists when they discovered the bones of extinct cave bears. And since systematic excavations started there in the early 1990s, the site has produced hundreds of hominid fossils, representing the remains of at least 28 individuals of both sexes. Although the individual bones are typically badly broken, the preservation of the bone itself is remarkable, and scientists have been able to reassemble the many hundreds of pieces into half a dozen more or less complete crania and numerous elements of the postcranial skeleton. A homogeneous sample this large of a single extinct hominid species from one place is unprecedented; and the Sima gives us a unique glimpse into the biology and even the demography of an extinct hominid species. The individuals whose bones were found at the bottom of the pit range from a single child to a handful of older adults in the 35- to 40-year range, and half of them had died between the ages of 10 and 18. Presumed males were larger than presumed females to about the same degree you find in modern humans, and one male stood almost six feet tall.

These were heavily built folk, with robust bones, and each probably weighed a good bit more than a modern human of the same height. Almost certainly, they were immensely strong compared to us. At the same time, their brains were on average a little smaller than ours, three crania ranging in volume from 1,125 cc to 1,390 cc. Their pelvises were broad, with birth canals capable of accepting the head of a modern newborn. Once individuals had emerged into the world, they faced a life reasonably free of dietary stress. Episodes of malnutrition are reflected in the enamel of developing tooth crowns, and evidence of such events is rarer at the Sima than it typically is in recent *Homo sapiens* populations. This is something not unexpected in an environment that is believed to have been rich and productive.

Studies of the mammal bones found at sites more or less contemporaneous with the Sima imply that northern Spain had cooled off a bit since Gran Dolina times, but that the Sima folk had lived in an open woodland landscape supporting a diverse fauna. The Atapuerca researchers believe that the Sima hominids would have been major predators in this setting, though they would also have been in competition with at least two species of lionlike large cats, recent arrivals in the region. The prevalence of arthritis in the Sima individuals' jaw joints coincides well with heavy wear on their teeth that indicates not only that they ate a fairly tough and abrasive diet, probably with a gritty plant component, but that they used their teeth extensively for such tasks as processing hides. And although the best-preserved skull shows evidence of a dental infection that may have (very painfully) killed its owner, many teeth in the sample show a concern for dental hygiene in the kind of grooving that comes with the frequent use of toothpicks.

The morphologies of both the skulls and the postcranial bones of the Sima hominids are like nothing known from anywhere else, although they do show a clear affinity to *Homo neanderthalensis*. Still, they were equally clearly not Neanderthals. In its morphology *Homo neanderthalensis* is a well-delineated species, with a large number of highly characteristic features in its skull. But not all of the distinctive Neanderthal traits are seen in the Sima hominids, though some are—for example, the thick ridges that arc delicately above each eye and a curious oval depression at the rear known as the "suprainiac fossa." The Sima folk remained less specialized, more ancestral, in such features as their steep-sided cranial vaults and relatively broad lower faces. They were certainly forerunners of the Neanderthals; but, befitting their early time frame, they were not the same thing.

It is not easy to date a pile of bones at the bottom of a pit, but fortunately lime-rich water flowing over the rubble pile at the Sima had laid down a limy cap over it soon after it had accumulated. And modern techniques make it possible to date such "flowstones," using radioactive isotopes of uranium deposited in the calcite crystals that form the stone. These unstable isotopes decay at a constant rate into stable isotopes of thorium that were not originally present, so the ratio between the two allows you to determine the time elapsed. High-precision measurements of both isotopes in the Sima flowstone have produced a series of dates clustering around 600 thousand years ago, with a minimum of 530 thousand. The possibility remains that the hominid bones are younger than this, as was at first believed; but either way, in terms of time they are well placed to be those of Neanderthal precursors.

So what was this jumble of fractured and disarticulated ancient individuals doing at the bottom of a deep, narrow shaft in a gloomy cave interior? This was certainly not a living place, and it is highly unlikely that 28 individual hominids fell in there by accident. Neither is there any suggestion that this could have been a carnivore den, although various carnivores did tumble in, including cave bears that may have become trapped while looking for a place to hibernate. Other carnivores may

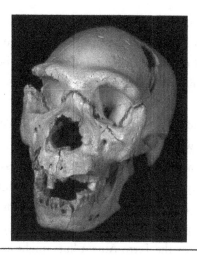

Skull 5 from the Sima de los Huesos at Atapuerca in Spain. Broken but restored, this is the best-preserved skull from a site that has yielded the fragmentary remains of at least 28 individuals some 600 thousand years old—the most amazing trove of hominid remains ever found. The population from which they came was a precursor to the Neanderthals. Photo by Ken Mowbray.

then have been attracted by the stench of their decaying bodies. But there is not one fossil of a browsing or a grazing mammal down there: this is anything but a random sampling of the local fauna. The Atapuerca researchers suggest that the hominids must have been deliberately thrown into the pit by their fellows, presumably as a way of disposing of them after they had died, somewhere outside the cave.

Not everyone is impressed by this explanation, but in its support the Atapuerca group point to one striking piece of evidence: the one and only artifact found in the pit happens to be a splendid handaxe made from rosy quartzite. Not only is this kind of artifact an unusual occurrence at any Atapuerca site of this age, but quartzite itself is also rare there. Early stone tool makers prized good raw materials, and, especially given the aesthetic appeal of the stone from which it was made, the Atapuerca group is almost certainly right in believing that "Excalibur" was a very special object to its possessor. Whether they are also right in believing that it was an overtly ceremonial object—it was apparently never used for any practical purpose—is more debatable. Even more hypothetical is the added deduction that this was a symbolic piece, tossed into the pit as part of a funerary rite. But if indeed it was such a thing, it would at the very least imply that the Sima hominids had developed a substantial sense of empathy, and it would certainly bolster the Spanish researchers' view that the Sima folk already possessed some power of symbolic thought.

Still, this is reading a great deal into one isolated observation, the true significance of which is entirely conjectural. Sadly, we have no other archaeological knowledge of the Sima people. No fossils like them have yet been found anywhere else, and we cannot confidently associate them with material expressions from (very rare) archaeological sites of their period in Europe (although it's not altogether impossible that the Schoeningen spears or the Terra Amata huts might have been the work of later members of their Neanderthal lineage, rather than of the contemporaneous *Homo heidelbergensis*).

The situation has been confused yet farther by the finders' allocation of the Sima fossils to *Homo heidelbergensis* (which they clearly are not), instead of to a new species affiliated with the Neanderthals (which *Homo heidelbergensis* is clearly not). But maybe there is an alternative approach to determining whether or not the Sima folk were symbolic, since their morphology leaves no doubt that they belong to a form antecedent to *Homo neanderthalensis*. The later Neanderthals left behind a rich archaeological record, one that furnishes us with a much firmer base on which to make such judgments. If the Neanderthals were symbolic, then the Sima hominids might have been. But if their successors the Neanderthals were not symbolic, then they weren't either.

TEN

WHO WERE THE NEANDERTHALS?

omo neanderthalensis occupies a very special place in the hominid pantheon because it was the first extinct hominid species to be discovered and named, back in the mid-nineteenth century. Largely as a result of this accident of history, the Neanderthals have always loomed very large in considerations of our own evolution—although it has for long been evident that they were not direct human precursors as was suggested early on, and there is fairly general agreement by now that they deserve recognition as a distinctive hominid species in their own right. This distinctiveness is evident in the fact that there is surprisingly little disagreement in the normally contentious paleoanthropological fraternity over which particular fossils are Neanderthal.

A braincase from a site in the north of France known as Biache-St-Vaast represents the earliest distinctively Neanderthal fossil. It dates from at least 170 thousand years ago (MIS 6), and the accompanying fauna indicates that conditions then were moderately cold. If you want to push the oldest Neanderthal occurrence back a bit farther, you might include a somewhat less complete braincase from the German site of Reilingen that is uncertainly dated to MIS 8, perhaps 250 thousand years ago. This is about the presumed age of another, more complete specimen from

Steinheim, also in Germany, that possesses more Neanderthal features than the Sima hominids but that, like them, is not fully Neanderthal. These tantalizing observations hint that events in the hominid history of Europe around this time were more complex than has generally been assumed, and it also suggests that we are never likely to find full-fledged Neanderthal fossils at more than about a quarter of a million years ago. Nonetheless, it's obvious that the Neanderthal lineage must have been present in Europe between Sima and Reilingen times, and it's possible that we know so little about it due to the effects of repeated glaciation and deglaciation in the region.

One of the reasons why we have such a good hominid record in Europe is the extensive occurrence of limestone rocks offering caves and overhangs that hominids would have been eager to exploit for shelter. The occupation debris they left behind in such places would regularly have been washed out by the water that flooded across the landscape each time the ice sheets melted; but the record is good enough to tell us that *Homo heidelbergensis* also existed in Europe during the tenure of the Neanderthal lineage. This knowledge strongly supports the idea that a complex minuet among hominid species was unfolding in Europe during the Middle Pleistocene (the period between about 780 thousand and 126 thousand years ago). If so, the large-brained Neanderthals were the victors in this particular contest, since by Biache times, if not well before, they were in sole occupation of the subcontinent.

In their 200-thousand-year tenure, the Neanderthals spread widely in Europe, and far into western Asia. Their fossils have been found as far south as Gibraltar and Israel, and what is reasonably an early Neanderthal archaeological site, dating from a warm interlude, has been found as far north as Finland. A recent report even places these hominids (by tools they are assumed to have made, rather than by their fossils) at a site in northern Russia not far from the Arctic Circle, at some 31 to 34 thousand years ago when conditions were considerably colder. In the west, Neanderthal fossils are known from north Wales in the British Isles, and numerous others are scattered eastward as far as Uzbekistan. A nondescript bone bearing the characteristic Neanderthal genetic signature has even been discovered farther east yet, at a site in the Altai Mountains of southern Siberia.

Neanderthal sites are thus spread over a vast area of the Earth's surface, and occur at a huge variety of altitudes, topographies, and latitudes. It is, then, clear from its distribution alone that *Homo neanderthalensis* was a rugged and adaptive species, able to cope with a wide array of different environments. Still, Neanderthals notably tended to avoid areas that were uncomfortably close to the glacial fronts, and the total area within the enormous range that they were able to occupy at particular points in time must have varied widely amid the climatic vagaries of the Pleistocene. For example, during a cold snap during about 70 to 60 thousand years ago the Neanderthals seem to have been limited to Europe's Mediterranean fringes, while during the warmest parts of the following MIS 3 their traces are found far up into northern and central Europe.

This is particularly interesting since it has long been assumed, on the basis of their northerly Ice Age origins, that the Neanderthals were somehow "cold-adapted." In sharp contrast to the African-derived and "tropically adapted" *Homo sapiens,* they were seen as creatures of the ice and snow. In reality there is very little to suggest this, either in the peculiar form of the Neanderthal nasal region, often interpreted as a mechanism for warming and humidifying cold and dry incoming air before it hit the fragile lungs, or in their limb proportions. These were long taken as adapted for the Arctic, but they actually seem to resemble what is seen in intensive modern foragers of varying backgrounds. The reality is that, over their long tenure, the Neanderthals ranged throughout many diverse territories and climates, to which they must have accommodated culturally. Indeed, it would have been impossible for them to accommodate otherwise, since it's been calculated that, under the coldest conditions these hominids endured, a 180-pound Neanderthal would have required an extra 110 pounds of subcutaneous fat to compensate for a lack of clothing. Being built like a Sumo wrestler is hardly what you might view as the ideal adaptation to a hunting lifestyle; and it is far more likely that the Neanderthals were as lean as Arctic peoples tend to be today, and depended on clothing and other cultural accoutrements for insulation and warmth.

Interestingly, analysis of two Neanderthal DNA samples (more on Neanderthal DNA in a moment) suggested that they possessed an inactive

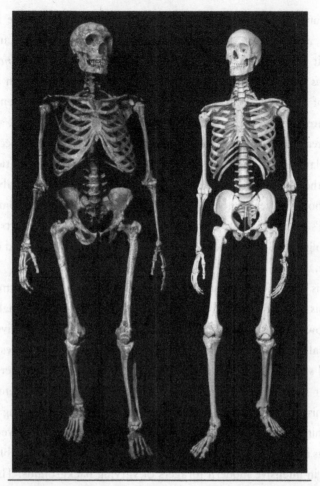

A reconstructed Neanderthal skeleton (left) compared with that of a modern human of similar stature. The comparison reveals two distinctly contrasting hominids. Apart from the cranial differences, note especially the very different shapes of the thoracic and pelvic regions. Photo by Ken Mowbray.

version of a gene affecting skin and hair color. Apparently, befitting their temperate origin, these individuals would have possessed pale skins and red hair. But, significantly, the gene variant in question is not one that is found among modern humans, even redheads. This one observation is by itself emblematic of the fact that we need to get away from seeing Neanderthals as a less successful version of ourselves, a hyper-adapted variety of modern human that put all of its eggs into the wrong basket.

Just as members of our species do, Neanderthals varied a bit in appearance from individual to individual, from place to place, and from time to time. Still, again like us, they all shared a distinctive common physical aspect. While Neanderthal braincases were capacious, they were also long and low, bulging at the sides and protruding at the back. (In contrast, our skulls are lightly built and globular, with a tiny face tucked underneath the front of a high, balloon-like braincase.) Neanderthal faces, hafted somewhat in front of the cranial vault, bore large noses (within which were some very unusual bony structures), and their cheekbones retreated rapidly at the sides. Below the neck the contrasts were equally striking. Compared to us the Neanderthals were heavily built, with thick-walled long bones bearing large, clunky joint surfaces at each end. Where our torsos are formed rather like barrels, tapering inward at top and bottom, theirs were funnel-shaped, tapering outward and down from a narrow top to match a broad, flaring pelvis below. This evidence of the skeleton thus joins other details in favoring the notion that Neanderthal gait was different from ours, stiffer and featuring greater rotation of the hips during striding. Beyond this, the general robustness of the Neanderthal skeleton also suggests great strength, and perhaps also high metabolic demands. Altogether, we are looking at a hominid that, although a fairly close relative, was anatomically distinct from *Homo sapiens* in numerous important details—although it seems to be us, not them, who have departed from the general hominid pattern in acquiring our unusual slender and gracile build. As far as we can tell from a less-than-perfect postcranial fossil record (though one that, significantly, contains the wonderful Sima sample) a broad pelvis and robust bone structure seem to have been characteristic of the entire Neanderthal lineage, and probably of all early *Homo* species.

We also differ from Neanderthals—and, as far as we know, from all other hominids—in the way in which we achieve our adult body form. We saw earlier that the Turkana Boy and other individuals of the *Homo ergaster/erectus* grade appeared to have developed much faster than *Homo sapiens* does, resulting in much shorter periods of both dependence and learning. And, despite its large brain, *Homo neanderthalensis* was no exception to this pattern. A recent study of Neanderthal dental development, using ultra-high-resolution techniques, has revealed that while the Neanderthal developmental period was indeed extended relative to earlier hominids, it was nonetheless shorter than our own. For example, the upper wisdom teeth (third molars) of one Neanderthal began developing at under six years of age, which is between three and four years earlier than in modern human children. Similarly, the first molars erupted substantially earlier in Neanderthals than in us. Translated into the overall developmental schedule, such data imply strongly that Neanderthals had a significantly shorter period of dependence on their parents than we do, and followed a faster path to sexual maturity. This conclusion coincides with analyses of the Neanderthal genome, which reveal that genes relating both to bodily and cognitive development differ from their equivalents in our own genomes.

Neanderthals also attained their characteristic cranial form through developmental trajectories that were not only faster than ours, but distinctly different. Sophisticated imaging and modeling techniques have shown that many of the characteristics that differentiate our faces from those of Neanderthals not only follow distinctive pathways of development postnatally, but also are well established at the time of birth. We cannot regard those many differences as superficial. Yet the actual shape of the brain is not among those features that are distinctive early on. Like Neanderthals, humans are born with longish skulls, which turns out to be a requisite of getting the neonate successfully through the birth canal; and we achieve our globular braincases in the first year of life, in the very rapid developmental spurt that propels the brain toward its unique external form. This dramatic early alteration in external form of the modern human brain and braincase is very unusual; and it is only possible once the constraints of the birth process are relaxed. The scientists who discovered it speculate that it may in some way be related to the internal reorganization of the brain that makes symbolic cognition possible.

NEANDERTHAL GENES

In 1997 *Homo neanderthalensis* became the first extinct hominid species to have its DNA characterized. In that year, a German team ingeniously extracted a length of mitochondrial DNA (mtDNA) from the original Neanderthal specimen that had been found in Germany's Neander Valley in 1856. Mitochondrial DNA is a short ring of DNA that resides in tiny organelles that supply energy to each of our cells. These contain their DNA independently of the much greater quantities of the stuff in the cells' nuclei—and this is a huge advantage for scientists trying to compare the mutations that have accumulated over evolutionary time. The advantage arises from the fact that mtDNA is inherited uniquely through the mother and, unlike nuclear DNA, doesn't get jumbled up in each generation as the egg and sperm of the parents combine. The historical message it contains is thus much simpler to sort out. Among modern humans, mtDNA has turned out to be an amazingly useful marker for characterizing various populations and tracing their spread; and the Neanderthal mtDNA turned out to fall well outside the envelope of variation that describes all human populations today. To be precise, while the German researchers found an average of eight differences in the relevant part of the mitochondrial genome between pairs of modern human populations, and about 55 between humans and chimpanzees, the number for Neanderthals was 26. What's more, the Neanderthals lay equidistant from all the modern human populations tested.

Since 1997 mtDNA has been obtained from numerous Neanderthal specimens originating in all parts of the species' range, always with the same result. As expected, the Neanderthals differed somewhat among themselves, though a relatively low diversity has suggested to researchers that Neanderthal populations were typically small, something that archaeologists had also guessed from the low relative abundance of the occupation sites they left behind. All the Neanderthals still clustered together, in contradistinction to *Homo sapiens,* and in numerous studies researchers have been unable to detect any Neanderthal contribution to the DNA of an extensive sample of modern Europeans.

These findings reinforced the notion derived from anatomical studies that *Homo neanderthalensis* was its own species, an effectively individuated entity with its own history and its own fate. However, nature is an untidy place, and species can be leaky vessels—especially where they are very closely related as well as actors in a fast-moving evolutionary drama, as hominids during the Pleistocene most assuredly were. In 2010, the German group announced another first—a draft version of a complete Neanderthal nuclear genome (taken from three samples of bone from the Croatian cave of Vindija, dated to about 40 thousand years ago). These samples provided a vast data base. There are more than three billion individual "nucleotides"—basically, data points—in a human genome; and interpreting this Neanderthal genome meant massaging all of those data points through some very hefty computer algorithms. But after all the necessary manipulations (with which not everyone is entirely happy), the researchers reported that "Neandertals shared more genetic variants with present-day humans in Eurasia than with present-day humans in sub-Saharan Africa, suggesting that gene flows from Neandertals into the ancestors of non-Africans occurred before the divergence of Eurasian groups from each other." Actually, on closer examination the apparent gene flow (i.e., gene transfer due to interbreeding) turned out to be in the order of 1 to 4 percent: hardly vast and, oddly, only one way: from Neanderthals into modern humans.

Even odder is a result reported by the same group shortly thereafter. These industrious researchers had already found that a morphologically uncategorizable finger bone from the southern Siberian cave of Denisova, only some 30 thousand years old, yielded a DNA fingerprint that distinguished it from both modern humans and Neanderthals, although it seemed to be somehow related to the latter. A complete genome was then obtained from this specimen, and is said to share a small proportion of its genes with modern-day Melanesians (and nobody else), suggesting—if true—that the ancestral Melanesians might have picked up these genetic variants on their way out of Africa and across Asia to the Pacific. A molar from Denisova produced basically the same genetic signature; but this tooth is both extremely large and morphologically dissimilar to any other hominid teeth known from so late in time, emphasizing that morphological and genetic evidence may sometimes be apparently at odds. However these findings are eventually interpreted, they suggest that events in later hominid evolution may have been very complex, and that the historically and functionally individuated entities that we recognize as hominid species may nonetheless have occasionally exchanged genetic material.

Perhaps such an exchange has even been an important source of genetic innovation in the human past. Not long ago, a group of molecular biologists in Chicago reported that a rapidly spreading variant of the microcephalin gene, important in regulating brain size, appeared to have been imported into the *Homo sapiens* genome only some 37 thousand years before the present. Their calculations suggested that it might have been introduced into our species from a relative that had separated from our lineage a little over a million years ago; and the Neanderthals seemed to them to fit the bill, though in fact any other hominid "donor" species might have been involved. At this point it's probably too early to know quite what to make of observations such as these (and 37 thousand years ago is too late to have made a material difference in the emergence of our own species); but it is not out of the question that minor gene exchange among closely related hominid species at an earlier time may have had a significant role in furnishing the ancestors of *Homo sapiens* with new genetic material.

This in itself is nothing remarkable. It has long been known that genes are occasionally exchanged between well-differentiated mammals. Indeed, there is a pair of ligers—huge hybrid beasts with lion and tiger parents—resident in an animal park in South Carolina right now. These are fearsome creatures indeed; and especially in view of their vigor you might be surprised to learn that lions and tigers are not even each other's recently diverged closest relatives. Lions are actually more closely related to jaguars, and tigers to snow leopards; and the last common ancestor of lions and tigers lived around four million years ago. But in spite of these impressive hybrids, nobody is out there arguing that lions and tigers are not fully individuated entities, each one with its own independent history and evolutionary trajectory. Despite that little genetic romp, there is no reasonable likelihood whatever that the two big cats will ever merge into a blended unit combining the characteristics of both parental populations. Closer to hominid home, the same thing seems to go for closely related primates that intermix. In Ethiopia, hybridization regularly occurs in a specific zone between hamadryas and gelada baboons, two closely related monkeys that are strikingly different to the eye. But even there, we see no indication that either broader parental species is losing its distinctive physical identity.

To put all this in context, the difference in skull structure between *Homo neanderthalensis* and *Homo sapiens* is far greater than what we see between hamadryas and geladas—and also greater than the one between lions and tigers. And whether or not acts of mating may occasionally have occurred between members of the two hominid species, the probability is negligible that there was any evolutionarily significant genetic interchange between them. In other words, nothing seems to have occurred that might have influenced the future fate of either, and the populations never integrated to any significant extent. Claimed "hybrids" such as the very late skeleton discovered at the Abrigo do Lagar Velho in Portugal, or the odd early *Homo sapiens* skull from the Peştera cu Oase in Romania, turn out on closer inspection to be somewhat unusual modern humans. What's more, and very significantly, the archaeological record is in parallel sending us a more or less identical signal of inconsequential or nonexistent cultural intermixing. From every line of evidence we have, it seems that *Homo sapiens* and *Homo neanderthalensis* were differentiated entities, each with its own history and way of doing business. Even if the odds may be reasonable that there was occasionally a bit of Pleistocene hanky-panky, swapping the odd stretch of DNA didn't change that functional reality.

NEANDERTHAL DIETS

As we've seen, the genetic evidence hints that Neanderthals were always thin on the ground, and this is probably also reflected in the typically small size of the sites they left behind, as well as their low density. In both warmer periods and cooler ones, Neanderthals lived in seasonal environments that would not have been enormously productive of the kinds of plant foods necessary to sustain hominids; and at all times they would likely have been quite heavily dependent on animal fats and proteins to get by. Just how dependent clearly varied, though, and this variation seems to have been largely a function of time and circumstances, for Neanderthals were flexible foragers who knew how to exploit whatever the environment offered.

One study of adjacent occupation sites in western Italy was eloquent in this regard. About 120 thousand years ago, during a warm period (MIS 5e), Neanderthal occupations were brief, and the animal remains associated with them consisted mainly of the cranial remains of older individuals. The researchers concluded that the hominids at the site had scavenged the remains of animals that had died of natural causes: the heads were the last bits available when large carnivores had had their fill. In contrast, at 50 thousand years ago, when (coincidentally or not) conditions were much colder, the animal remains were those of individuals in the prime of life, and consisted of parts from all over the body. Together with greater densities of stone tools this suggested not only more sustained site occupation, but that the Neanderthals were using sophisticated ambush-hunting techniques to obtain carcasses that were brought back whole to be butchered at the home site. Archaeological evidence of this kind almost always gives an incomplete impression of the lives of the ancient hominids who left it behind, and it is never easy to interpret. Nonetheless, the contrast between the earlier and later occupations is striking; and at the very least it indicates not only that Neanderthal techniques of obtaining animal foods varied greatly, but also that their occupation habits did, too. These hominids were certainly not stereotyped in their subsistence strategies.

Flexible they may have been, but a powerful consensus is growing among archaeologists that under appropriate circumstances Neanderthals were top predators. Not only were animal products the main if not the only potential hominid mainstay at cooler times, but evidence is also accumulating that they routinely went after large-bodied mammals, some of them the most fearsome of all the creatures on the landscape. The most provocative such evidence comes from the study of stable isotope ratios preserved in Neanderthal teeth and bones. We've seen that carbon isotopes have been very informative about diet among the australopiths; in the case of the Neanderthals, an equivalent role has been filled by stable isotopes of nitrogen. It turns out that the ratio between the two isotopes ^{15}N and ^{14}N increases slightly in your tissues with every step you take up the food chain: the higher the ratio, the more meat there is in your diet. Starting in the early 1990s, scientists discovered that the bones of Neanderthals invariably showed higher ^{15}N/^{14}N ratios than were found in the fossil bones of herbivores from the same place;

indeed, they were right up there with the ratios recovered from wolves, lions, and hyenas—if not higher yet.

This observation fit well with the abundance of butchered herbivore remains typically found at Neanderthal sites. But the ultimate observation came in 2005, when a French team found an extremely high $^{15}N/^{14}N$ ratio in the bones of a very late Neanderthal from a place called St.-Césaire. Since this value was well above what they had found even in hyenas from the same site, the scientists suggested that the only way in which Neanderthals could possibly have achieved such a high ratio was by specializing in the consumption of herbivores that were themselves enriched in ^{15}N. And the only putative victims were among the most intimidating of the many large beasts roaming the landscape: namely, mammoths and wooly rhinoceroses. What is more, the French scientists suggested that it would not have been possible for the St.-Césaire Neanderthals to have scavenged all the mammoth and rhino carcasses that would have been necessary to sustain the high nitrogen isotope ratios they had found in the hominids' bones. In their view, the hominids must have actively hunted the huge mammals, presumably as an important component of a long-standing dietary tradition. The case seems pretty strong, then, that Neanderthals were redoubtable hunters who, even at low population densities, were able to tackle some of the most formidable prey around. At their living sites they routinely controlled fire in hearths; and these fires doubtless provided a focus of their social activities, besides furnishing a means for cooking all that meat and for discouraging unwanted predators.

Still, it's important not to forget that plant foods must have played a significant role in the Neanderthals' diets in most places and at most times. This aspect of their food intake has been predictably neglected because plant remains rot rapidly, and rarely preserve in the archaeological record. However, scientific ingenuity is beginning to open up some amazing new avenues for investigation. For example, a recent report describes plant microfossils (both starch grains and phytoliths, tiny rigid bodies that occur in plant roots, leaves, and stems, and differ according to plant species) that were recovered from the plaque coating Neanderthal teeth from two famous sites. A dentist's nightmare had become a treasure trove for paleoanthropologists. One of the sites in question is the cave of Shanidar in northern Iraq, and the specimen examined dates from about 46 thousand years ago. Shanidar is, by the way, the site that has famously yielded the skeleton of an aged male Neanderthal with a withered arm. This appendage must have been useless to its possessor for most of his long life, and his survival has elicited speculation that he enjoyed the sustained support of his social group. The other site is the Belgian cave of Spy which, at about ten thousand years younger, falls very late in Neanderthal history.

Though far apart in time and space, and representing environments ranging from Mediterranean to cool temperate, the two caves tell similar stories. In both places the Neanderthals consumed a wide variety of plant foods that reflected the range of resources available in the local environment. There was no indication of specialization on particular plants, but in both places many of the foods would have required some preparation prior to consumption, and some starchy plant parts had indeed been cooked to render them more edible. There is, by the way, no contradiction between extensive consumption of starches and the nitrogen isotope record, because the isotopes only register the consumption of meat and of plant foods that are high in protein. At Shanidar the foods indicated by the microfossils include dates, barley, and legumes—items that would have been ready for harvesting at different times of year, thus indicating that foraging for plant foods was a year-round activity. All in all, this new study shows us that the essentials of the modern hunting-gathering style of subsistence had been established by the time the Neanderthals had entered the picture. Like *Homo sapiens* today, *Homo neanderthalensis* was an opportunistic omnivore, reminding us that despite our secondary adoption of a predatory lifestyle, we have never entirely put behind us our ancient vegetarian heritage.

NEANDERTHAL LIFESTYLES

Apart from being small, we didn't know until very recently what those Neanderthal groups that sat around the fire cooking their food were actually like. All we had as a basis to speculate on the subject were stone artifacts and broken bones, and the ways in which these were scattered around living sites. This scattering was typically (though not invariably) random, with little suggestion that the living space was divided into areas for specific activities such as butchery, stone knapping, sleeping, eating, and so forth. We routinely find such division of space at sites left by fully symbolic modern humans, so there is already some suggestion of different approaches to domestic life by the two species. But until recently, there hasn't been much to tell us how Neanderthal groups were organized. Now a team of Spanish researchers, working at the 50-thousand year-old Neanderthal site of El Sidrón, has come up with some intriguing suggestions based on both physical and molecular evidence.

The El Sidrón site itself is a long and complex warren of tunnels produced in the surrounding limestone by an ancient underground river system, and it has a complex history. Most notably, an extensive assemblage of Neanderthal bones was deposited in a single event on the bottom of one arm of the cave, when the ground surface above (or, just possibly, the floor of a higher tunnel) collapsed into the cavity below. Large numbers of knapped stones were intermixed with fossil bones and other debris. Many of the fragments could be refitted into complete cobbles, suggesting that the spot where the collapse occurred was a place where stone tools were made. The 1,800 fossil fragments found in the debris represent the broken-up remains of twelve Neanderthal individuals: six adults, three adolescents, two juveniles, and an infant. All appear to have already been dead when the collapse occurred, not long after their decease. More remarkably, not only had these Neanderthals been dead, but the researchers conclude that they had been the victims of a massacre, since many of the bones show marks of cutting and percussion consistent with defleshing, and probably cannibalism.

Evidence of defleshing is not uncommon on Neanderthal (and even *Homo heidelbergensis*) bones, and many scientists have argued that removal of flesh from corpses after death is not necessarily proof of cannibalism; but the case made that the hominid bones at El Sidrón were broken for consumption is a compelling one, and the probability seems to be growing that this behavior was indeed part of the Neanderthal repertoire. Interestingly, the El Sidrón researchers think that, in contrast to the "gastronomic cannibalism" seen at the Gran Dolina (i.e., cannibalism occasioned by habit, rather than by necessity), the El Sidrón

Neanderthals were the victims of "survival cannibalism." In support of this they point to the fact that the fossil remains bear clear signs of environmental stress, mainly in the form of an abundance of those defects in dental enamel formation that were notably rare at the Sima de los Huesos. If dietary stress was indeed a significant issue for these hominids, then it is likely that competition among contiguous Neanderthal groups for available resources was strong. Putting the various lines of evidence together, the researchers conclude that the twelve El Sidrón Neanderthals all belonged to a single social group that had been ambushed, killed, and consumed by another.

Two further observations support the notion that an entire Neanderthal group had perished in the El Sidrón event. One of these is that a group size of twelve, with a few adults of each sex and children of all ages, is pretty much in line with what you might expect. Specific estimates of Neanderthal group sizes are few and far between, but one recent study at the 55-thousand-year-old Spanish Neanderthal site of Abric Romaní concluded that groups occupying the rock shelter had varied in size from eight to ten individuals. If the Abric Romaní inhabitants were typical, and the estimates of their group sizes are accurate, it's even possible that the twelve individuals from El Sidrón belonged to a largish social unit by Neanderthal standards.

Still, wherever this band stood in the size spectrum, the notion that it constituted a single social unit was supported by analysis of its members' mtDNA, which had been excellently preserved in the cool conditions within the cave. For a start, diversity among the El Sidrón mtDNA genomes was very low, consistent with a family group. But most revealing was the finding that the three El Sidrón adult males had all belonged to the same mtDNA lineage, while each of the females had belonged to a different one. And here, for the first time, is a potential (though not definitive) message about the social organization of Neanderthals: that the El Sidrón males had remained in their birth group, while the females had married out of theirs, being dispatched at or soon after puberty to join a neighboring band. As one scientific colleague was quoted by the *New York Times* as saying, "I cannot help but suppose that Neanderthal girls wept as bitterly as modern girls, faced by the prospect of leaving close family on their 'wedding' day." This may be anthropomorphizing a bit—and it is certainly true that impassive female transfer is not that uncommon among primates—but it is difficult not to respond to the sentiment.

The inferences made by the El Sidrón researchers about Neanderthal society do not stop there. They note that a five- to six-year-old child and a three- to four-year-old were probably offspring of the same adult female. This suggests a birth interval of around three years, consistent with what was historically seen among hunter-gathering peoples. This in turn implies that Neanderthals achieved prolonged inhibition of ovulation, most plausibly through the expedient of protracted breastfeeding. An imaginative further conjecture comes from the material from which the El Sidrón stone tools were made: the nearest place at which it could be obtained was several miles away. Perhaps, the researchers speculated, the El Sidrón Neanderthals had incurred the wrath of the neighboring group into whose territory they had forayed to obtain it, and paid a heavy price in a reprisal raid.

Taken together, all of this tantalizing evidence from El Sidrón is helping create a more visceral picture of the Neanderthals than we ever had before. Knowing from high-tech laboratory analyses that tiny numbers of Neanderthals heroically hunted mammoths out on the tundra certainly evokes our admiration of these hardy and resourceful hominids. But this kind of information is profoundly different from contemplating the historical vignette of Neanderthal life—and death—with which El Sidrón presents us. The vision of a peacefully stone-knapping extended family of Neanderthals being raided, murdered, butchered, and eaten by a marauding group of their fellows is an unsettling one in the extreme; but then again, it is possibly not so different from what every modern watcher of crime-scene television is by now inured to.

On the more humane side, one of the reasons we have such a good sampling of reasonably intact Neanderthal remains is that these hominids at least occasionally buried their dead. And while it has been argued both that the presumed burials never occurred, and that they not only occurred but sometimes contained grave goods, the truth seems to lie somewhere in between. Yes, the Neanderthals did invent the practice of burial; and no, there is no really convincing evidence that they ever did so with the ritual that typically accompanies modern human burials. Much as we want to see echoes of ourselves in this practice (which Neanderthals apparently invented before our ancestors did), it is impossible to know whether or not Neanderthal burials were overlarded with all of the symbolic baggage with which ours are. That they imply some sort of deep empathetic feeling seems close to certain; but in the broader context of what we know about Neanderthals, it is far less probable that they imply belief in an afterlife—something that would indeed demand symbolic cognitive abilities.

NEANDERTHALS AND MATERIALS

By the time diagnostic Neanderthal remains are known in Europe, the stone-working tradition known as the "Mousterian," using variants of the prepared-core technique, had become entrenched. Indeed, in Europe the Mousterian is virtually synonymous with *Homo neanderthalensis*, although a very similar toolkit was also produced by other hominids in North Africa and the Levant. The most characteristic implements of the Mousterian are modestly sized sharp points and convex-sided scrapers, or even small teardrop handaxes made on flakes; but the number of variations is endless. This may not, however, have been through the toolmakers' specific intention. For while more than 50 distinct Mousterian tool forms were defined by mid-twentieth-century archaeologists, more recent researchers have recognized that there is in fact more of a continuum of form. This is due to a complex and discontinuous sequence of actions, as flakes made from superior materials were continually resharpened to maintain their functionality. Indeed, it was cleanly and predictably fracturing rocks themselves that were the key to making the best Mousterian tools. Good materials were evidently highly prized and regularly sought far afield, showing how valuable they were. Not infrequently, the nearest source of the rock used to make at least some of the tools found at Mousterian sites was many miles away—hence the speculation over the fate of those unfortunate Neanderthals at El Sidrón.

The need for good materials was occasioned by the Mousterians' sheer skill, for they were gifted stoneworkers who disdained poor materials, only making crude implements out of them when—as was frequently the case—there was no alternative. The Neanderthals instinctively knew

Mousterian flint tools made by Neanderthals at various sites in France. These skillfully shaped tools include two small handaxes, two scrapers, and a point, all made on stone flakes using the prepared-core approach. Photo by Ian Tattersall.

stone, as a modern cabinetmaker instinctively knows wood. And while a piece of silicified limestone might be good enough for producing a simple flake meant to be used only until its edge went blunt, the Mousterians carefully fashioned a good piece of flint or chert, then gave it a new edge over and over again until it was too small to be of further use. The discovery of scraping tools or points bearing traces of resin confirms that Neanderthals often set such tools into wooden handles, or used them as spear tips, binding them in position with leather thongs or sinews. The Mousterian toolkit was clearly the product of intelligent and dexterous beings.

Yet perhaps not beings just like us. Despite their frequent beauty, and for all the skill that went into making them, Mousterian tools showed a certain monotony over all the vast area that the Neanderthals inhabited. Several varieties of the Mousterian have been named, and are still recognized. But uniformity in concept was the rule of the day, and it's likely that the minor variations we do see in Neanderthal toolkits broadly reflect local differences in activity due to differential availability of resources, or occasionally to some refinement over time, rather than to the experimentation with different ways of doing things you'd expect to find among geographically scattered modern people. What's more, while they hafted stone tools into wood, Neanderthals rarely seem to have made tools of other soft materials. Bone and antler are plentiful at Neanderthal sites, and were abundantly fashioned into artifacts by later Europeans. But the Mousterian toolmakers rarely took advantage of these materials—although one of the rare examples of a Mousterian bone tool, from the 50-thousand-year-old site of La Quina and evidently used for the purpose of retouching stone tools, appears to have been made from a piece of hominid cranium. In this case and in others, the Mousterians bashed bones as though they were stones, with none of the sensitivity to the special mechanical properties of soft materials shown by their successors. In short, spectacular as it was, Neanderthal craftsmanship was pretty stereotyped.

The upshot of all of this is that we find nothing in the technological record of the Neanderthals to suggest that they were symbolic thinkers. Skillful, yes; complex, certainly. But not in the way that we are. As a species, *Homo neanderthalensis* seems to have fully participated in the hominid trend over time toward more challenging behaviors, and toward more subtle and intricate relationships with the environment. It certainly participated in the hominid trend toward bigger brains, possibly taking this tendency to its most extreme expression. But behaviorally there was no qualitative break with the past; the Neanderthals were simply doing what their predecessors had done, if apparently better. In other words, they were like their ancestors, only more so. We are not. We are symbolic.

ELEVEN

ARCHAIC AND MODERN

S tone implements and their means of manufacture are hardly iron-clad proxies for symbolic thought processes on the part of the toolmakers; and indeed it can be argued that we know of little if anything in Old Stone Age technology that could demonstrate such mental processes. Throughout this period, with few exceptions, we can confidently infer symbolic intent only from overtly symbolic objects, or from the results of explicitly symbolic actions. Of course, identifying such expressions is more easily said than done. Burial, as we have seen, may well have other motivations. And despite the fact that ochre was widely used in symbolic contexts by later people, there is no evident reason why the well-documented grinding of pigments that took place at various Neanderthal sites need necessarily imply intent of this kind. Even recognizing "symbolic" objects can be a tough call. A cave wall decorated with lively animal images leaves no doubts; but given that if you sufficiently desire to you may interpret a wide variety of scratches and other strange markings as symbolic, this can become a very gray area indeed.

With the Neanderthals we find ourselves at best somewhere toward the more dubious end of that gray area. And it must surely be signifi-cant that, from the entire expanse of time and space the Neanderthals inhabited, we have nothing that we can both confidently associate with them *and* unambiguously interpret as a piece demonstrating modern cognitive processes. There is certainly the odd straw in the wind, and a few uncertain objects are known that scientists argue about. But this is hardly unexpected in a record left behind by a big-brained close human relative that clearly displayed complex behavior patterns. What is almost certainly more telling than such putative flashes of the symbolic spirit, is that there is no substantive evidence that our style of thinking and its expression were routine aspects of Neanderthal consciousness or Neanderthal societies.

The most striking thing of all, though, is the astonishing contrast between the impressive but prosaic material record bequeathed us by the Neanderthals, and the symbol-drenched lives of the fully modern people who succeeded them in Europe. These new people, colloquially known as the Cro-Magnons, entered the subcontinent around 40 thousand years ago, bringing with them so-called Upper Paleolithic material cultures that, however distant from us, provide abundant evidence that these people viewed and experienced the world in essentially the same way that we do. Such evidence includes the astonishingly powerful art of the Lascaux, Chauvet, and Altamira caves that we encountered in

One of the world's earliest artworks: a carving of a horse in mammoth ivory, probably around 34 thousand years old. This is a supremely symbolic object: in its flowing lines it is not merely a representation of the chunky horses that roamed the Ice Age steppes of Europe, but an abstraction of the graceful essence of the horse. Vogelherd, Germany: drawing by Don McGranaghan.

chapter 1. And the Cro-Magnons' arrival on the Neanderthals' territory heralded an equally telling acceleration in the tempo of technological change, as the artists or their fellows explored the imaginative possibilities opened to them by the new form of reasoning. Clearly, for all the Neanderthals' formidable resourcefulness and skills, the Cro-Magnons were creatures of an entirely new order.

We see this not only in their material productions, but in less direct indicators such as the higher population densities that are reflected in the number and size of Cro-Magnon sites. Indeed, it is most probably the Cro-Magnons' abilities to exploit the environment much more intensively than the Neanderthals could, as well as their evident advantages in planning if it ever came to direct conflict, that led to the total disappearance of the latter within ten millennia of the new hominids' arrival. It has been argued that, in the run-up to the last peak of cold that occurred some 20 thousand years ago, Neanderthals were already in terminal decline; and this may well have been the case regionally, as in the southern tip of Iberia, which late Neanderthals seem to have abandoned before the Cro-Magnons arrived. But it is unlikely that no contact was made between the two kinds of hominid anywhere within the huge territory that the Neanderthals had inhabited, and there are some rather speculative indirect indications—in addition to the DNA—that the two species did encounter each other.

Cave entrances and rock overhangs are common features in limestone areas of Europe, and were preferred living places for early humans because of the natural shelter they provided. Still, the epithet "cave men" is certainly not justified. Neanderthals and Cro-Magnons alike roamed and camped widely over the landscape, and we associate them with caves simply because such places are relatively protected from erosion, and thus preferentially preserve the traces of ancient occupation. Many caves and rock shelters preserve multiple layers of debris left behind by successive generations of both Neanderthals and Cro-Magnons (usually as indicated by the artifacts they left behind—their bones are much more rarely found). Where a single site has evidence of both hominids, the Upper Paleolithic levels almost invariably overlie the latest Mousterian strata, the two most often distinctly separated by sterile sediments signifying that the site was abandoned for a period of time. Only two localities show possible evidence of Mousterian overlying Upper Paleolithic before finally being definitively replaced.

But at a few very late sites there is evidence of yet another cultural tradition—known as the "Châtelperronian" and found at a scattering of sites in western France and northern Spain—that incorporates features of both the Mousterian and the Aurignacian (the first cultural phase of the Upper Paleolithic). The Châtelperronian industry exhibits not only the "flake" tools of the Mousterians, but also "blade" tools like those that were a major feature of the Aurignacian tool kit, in addition to bone and ivory objects. As you'll recall, blades are those slender flakes, more than twice as long as wide, that occasionally also turn up in Africa in much earlier contexts; and in Europe they are a Cro-Magnon hallmark. In recent years, the Châtelperronian has generally been viewed as the handiwork of Neanderthals, possibly as a result of acculturation due to contact with modern humans, who were well established in Europe by Châtelperronian times. Sites attributed to the Châtelperronian all fall in the very brief 36-thousand- to 29-thousand-year range, whereas radiocarbon dates indicate that Cro-Magnons were already in Spain by 40 thousand years ago, having likely arrived from the east. It is worth noting, though, that radiocarbon dating in this remote time period is rather tricky, due partly to the minuscule amounts of radiocarbon that remain in samples of that age. Recent work indicates that dates obtained using older methods tend to be a bit young, and recent high-precision dates have suggested to some researchers that the period of overlap between the two hominid species was both earlier and briefer than traditionally believed—another reason for concluding that abrupt replacement was involved.

What is more, just what form any possible acculturation represented by the Châtelperronian may have taken is largely conjectural: suggestions as to how the odd combination of cultural features came about include trading, imitation, and theft. Still, some recent developments may have made all this moot, for the tide appears to be turning against the idea that the Châtelperronian bone and ivory items were the handiwork of Neanderthals—although the blade artifacts were clearly developed within an older Neanderthal tradition. Unquestionably the most famous potentially symbolic Châtelperronian pieces were found in a cave called the Grotte du Renne, at Arcy-sur-Cure, in France. They include a rather splendid polished ivory pendant that most people would have little difficulty in identifying as a symbolic object, and until recently they were believed to have been associated with some rather fragmentary Neanderthal fossils from the same site. But several independent studies have recently concluded that they were most likely introduced from above into the earlier Neanderthal layers, through the sort of natural mixing up of strata that can frequently occur in caves. Similarly, the association with the Châtelperronian of the clearly Neanderthal skeleton from St.-Césaire has been called into question by recent studies. The bottom line here is that, although it seems improbable that the Neanderthals and Cro-Magnons did not encounter each other once in a while, we still have no good record of any interaction between them, let alone of what form it might have taken.

Thus the obvious question—whether the large-brained Neanderthals *could* have acquired symbolic ways of dealing with information from the incoming Cro-Magnons—remains unanswerable on the basis of the material record we have to hand. But when we take all the indirect lines of evidence into account, it seems a bit unlikely. When Neanderthals and Cro-Magnons met on the landscape it seems probable that, for all their similarities, they would have perceived each other as alien beings, each with its own way of viewing and dealing with the world. Language would have been a major issue, among many others. Whereas the Cro-Magnons almost certainly possessed language as we know it today—however different their specific language might have been from any spoken now, or even within recorded history—it seems likely that the Neanderthals did not. Language is an intensely symbolic activity that, as we'll see in detail later, probably played a unique and pivotal role in the acquisition of modern symbolic consciousness; and even in the unlikely event that an occasional gifted Neanderthal managed to acquire its rudiments, there is no firm indication that any potential interchange had a material effect on the cultural or biological trajectory of either group.

When we ponder the differences between the Cro-Magnons—whose lives, like ours, were doubtless riddled with myth and superstition—and the Neanderthals, perhaps the closest thing we can obtain to a glimpse of the divergence in psyche comes from the grisly yet matter-of-fact fate of the hapless denizens of El Sidrón, and from the casual way in which that piece of skull bone from La Quina was used as the most inconsequential type of tool. I cannot help but read an intense form of focused practicality—and a related lack of symbolic imagination—into these and all the other material leavings of the Neanderthals. These large-brained relatives certainly were smart; but their particular kind of smartness was not ours. This difference is hard for us to comprehend fully. As I've already stressed, it is just not possible for a symbolically thinking modern human to project him- or herself into the mind of any creature that did not think that way—no matter how large-brained or closely related to us it might have been. The cognitive gulf is just too great. At our current stage of understanding we simply cannot know how Neanderthals subjectively experienced the world and communicated that experience to each other. All we can be certain of is that we do the Neanderthals a grave injustice by looking upon them as an unsuccessful version of ourselves.

TWELVE

ENIGMATIC ARRIVAL

About the same time that *Homo neanderthalensis* first appeared in Europe, our own species *Homo sapiens* was emerging in Africa. But while the Sima de los Huesos fossils give us a pretty good idea of Neanderthal ancestry in Europe, we have no African equivalent in our own case. A number of hominid crania are known from sites in eastern and southern Africa in the 400- to 200-thousand-year range, but none of them looks like a close antecedent of the anatomically distinctive *Homo sapiens.* Yet we can be confident that Africa was the continent of our birth, not only because the very earliest plausibly *Homo sapiens* fossils are found there, but because numerous DNA comparisons of modern human populations have made it clear that they all converge back to an African ancestry. The lack of anticipatory fossils might simply be due to the fact that Africa is a very large place that has not been explored in great detail; but it may also suggest that our unusual species originated in the kind of systemwide genetic regulatory event I have already mentioned in the case of the also radically new *Homo ergaster.* For *Homo sapiens* departs in numerous features from the much more ancestral body form exemplified both by the Neanderthals and those other extinct members of the genus *Homo* represented by relevant fossils. Still, this is not the whole story, for as far as *Homo sapiens* is concerned it appears that body form was one thing, while the symbolic cognitive system that distinguishes us so greatly from all other creatures was entirely another.

The two were not acquired at the same time, and the earliest anatomical *Homo sapiens* appear right now to have been cognitively indistinguishable from the Neanderthals and other contemporaries.

ANATOMICALLY MODERN *HOMO SAPIENS*

The first traces we have of people who looked in their bony structure exactly—or almost exactly—the way we do today, come from two sites in northeastern Africa. In the late 1960s, rocks in southern Ethiopia's Omo Basin that are now reckoned to be about 195 thousand years old yielded the fragmentary remains of a skull that, reconstructed, looks plausibly to be a *Homo sapiens,* even if not exactly like a member of any human population living today. And much more recently, deposits at Herto in northern Ethiopia produced a trio of crania, including a fairly complete child and adult, that are also best considered *Homo sapiens,* if once more differing from today's people in a few details. Certainly the adult shows the characteristic high, voluminous cranial vault, with a small face retracted beneath its front, which is so conspicuously unique to our species. The Herto fossils can be firmly dated to between 155 and 160 thousand years ago; so, between them, the Omo and Herto hominids demonstrate pretty clearly that the distinctive basic *Homo sapiens* cranial anatomy was established by about 200 to 160 thousand years ago. Importantly, this date range coincides with the dates for the origin of *Homo sapiens* proposed by molecular anthropologists, based on the time-to-coalescence calculated for a large number of different modern human populations from around the world.

Still, in cultural terms it seems that the reign on Earth of *Homo sapiens* started with more of a whimper than a bang. Stone tools found at both of the Ethiopian sites are unimpressive. At Omo the few artifacts found have been characterized as "nondescript," while at Herto handaxes are present, as well as prepared-core flakes. This is the latest recorded presence of handaxes in Africa, and it places the Herto stone tool assemblage right at the end of what was evidently a complex and prolonged transition from the Acheulean to the "Middle Stone Age" (MSA) technology associated with later humans. The MSA has regularly—if, as it turns out, a bit inappropriately—been regarded as the African equivalent of the Neanderthals' Mousterian in Europe, largely because both traditions relied upon prepared-core techniques. But, as we will shortly see, there seems to have been a lot more going on in the MSA than there ever was in the Mousterian—although as far as we know at present, these stirrings were not expressed until after Omo/Herto times.

Because of the persistence of the old alongside the new that has typified all of human technological history—and continues unabated today—it's hard to say exactly when the MSA began. But a general reckoning would place its origin in the period between about 300 and 200 thousand years ago, most likely before recognizable *Homo sapiens* had come into existence—and if so, comfortably in line with the disconnect we've already seen between biological and cultural innovation in human evolution.

The matter of *Homo sapiens* origins has been muddied by a long-running tendency among paleoanthropologists to identify reasonably large-brained non-Neanderthal hominids who didn't look like us as "archaic *Homo sapiens*." This appellation has been applied to specimens found in almost all regions of Africa, as well as elsewhere. But it really doesn't help much to include in our species creatures who didn't share at least the most basic aspects of our distinctive anatomy (most especially that reduced and retracted face). Among the most puzzling of such fossils are certain crania from North Africa, some of which are associated with a stoneworking industry known as the Aterian, after the site of Bir el Ater, in Algeria. The Aterian toolkit is fairly characterized as a variant of the MSA, but it includes some unique tool types, such as the defining "tanged points" that may have been hafted as spear tips or even conceivably, in very late phases, as arrowheads.

Long considered to be quite recent in date, the Aterian is now known to occur at some quite ancient sites, and this has excited speculation that the earliest producers of this industry may have played a role in the initial exodus of *Homo sapiens* out of Africa. Geographically this makes sense, for the Sahara desert has not always been the barrier to human movement that it is today, and areas that are now sandy wastelands have produced ample evidence of earlier occupation, notably by Aterians. As fossil drainage systems now covered by blowing sands testify, the Sahara has periodically "greened" as rainfall increased and lakes and vegetation sprouted everywhere. One of the wettest such periods occurred between about 130 and 120 thousand years ago (the time of the last interglacial in Europe). And at that point the Sahara could certainly have acted as a conduit for modern human populations expanding northward, although there are reasons for thinking that the Aterians themselves may have remained effectively in Africa, at least in the longer term.

One reason for uncertainty is the Aterian peoples' identity. Although it is generally unwise to associate any particular kind of hominid exclusively with a specific toolkit, it may be relevant that the North African hominid remains so far associated with early Aterian societies belong in that very dubious "archaic *Homo sapiens*" category. Best known among these fossils is a partial cranium, plus more fragmentary materials, from Dar-es-Soltan II, a site in Morocco that may be as much as 110 thousand years old or more. One that has received a lot of publicity lately is a crushed and fragmented but relatively complete child's cranium from Contrebandiers Cave, also in Morocco, which has been dated to about the same time and which, in its reconstructed form, is clearly not a standard-issue modern human despite a reasonably capacious braincase. Even more unlike modern humans are a couple of crania from another Moroccan site, Jebel Irhoud, which may be over 160 thousand

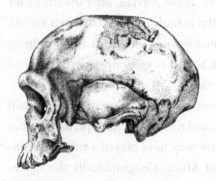
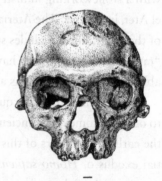

Front and side views of a cranium from Jebel Irhoud, Morocco, that is thought to be some 160 thousand years old. Often considered close to Homo sapiens, *it actually has a very distinctive facial structure. The associated industry is very much like that of the Neanderthals in Europe. Drawing by Don McGranaghan.*

years old. These earlier specimens are associated with a toolkit that is said to closely resemble the Mousterian of the Neanderthals—though the hominids themselves don't look Neanderthal at all. None of these North African specimens presents itself as an obvious variety of *Homo sapiens*, despite brain volumes for the Jebel Irhoud individuals of 1,305 cc and 1,400 cc. The more complete Jebel Irhoud 1 cranium has a rather small lower face; but the whole facial skeleton is forwardly positioned, and it also boasts prominent brows behind which the forehead retreats in a manner unlike anything we see in modern *Homo sapiens*.

Recognizing species from their bones is often a tough proposition among close relatives: in some cases, much physical diversity may accumulate within a population without speciation occurring, while in others, the bones of members of two species descended from the same ancestor may be virtually indistinguishable. In the absence of a good morphological yardstick we thus can't be absolutely sure that Aterians or the Jebel Irhoud people would not have been able to exchange genes with anatomically mainstream *Homo sapiens*. Indeed, they may conceivably have done so, as we'll see in a moment. Still, as we will also see, although Aterians could have ventured a bit beyond Africa at a very early stage, they certainly did not play a role in the definitive *Homo sapiens* exodus that later populated the world.

Now let's shift the scene to the nearby Levant, the area along the eastern edge of the Mediterranean Sea that would have been the first stop along the route followed by hominids leaving Africa and spreading north and west. Because the Levant has typically shared elements of its fauna with Africa rather than with regions to its north, biogeographers have actually often considered this area an extension of the great African landmass. Following Omo/Herto times we have a growing inventory of African fossil hominid crania that unequivocally show the distinctive modern *Homo sapiens* morphology; but none of them is indisputably as early as a more or less complete skeleton that was buried, by the latest reckoning 100 thousand years ago or more, at the Israeli cave site of Jebel Qafzeh. This specimen clearly represents a member of our own species, as do the remains of an adolescent found nearby. Yet at the same site we also find—in larger numbers—the remains of big-brained hominids who are not your standard-issue *Homo sapiens*, although they

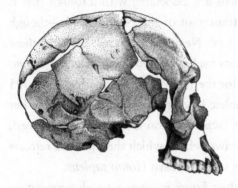
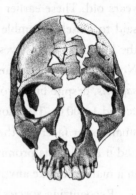

The earliest modern human fossil known from outside Africa, this skull was found at the Israeli site of Jebel Qafzeh. Known as Qafzeh 9, it is anatomically a standard modern human. However, other hominid fossils from the same site do not have typically modern skull anatomy, and all the Qafzeh hominids are associated with a Mousterian industry similar to made by Neanderthals in the same region. Drawing by Don McGranaghan.

are certainly not Neanderthals. To deepen the mystery, all of these hominids were found in association with Mousterian stone tool kits. These tools were more or less identical to those produced by the Neanderthals whose presence is well documented in Israel at around the same time. Indeed, Neanderthal sites in the region date from at least 160 thousand years ago to about 45 thousand years ago.

The Qafzeh hominids are often spoken of in the same breath as those from the rock shelter of Mugharet-es-Skhūl, a burial locality a few dozen miles away on the western slope of Mount Carmel, overlooking the Mediterranean. Excavations at Skhūl produced the remains of ten adults and juveniles, probably around 100 thousand years old. These fossils present a more uniform physical aspect than those from Qafzeh, but they are no less odd for that. Like modern humans they have high, rounded cranial vaults that held impressively large brains of some 1,450 to 1,590 cc; but, unlike ours, the fairly heavily built faces of the Skhūl hominids were unretracted, jutting proudly afore the vault and topped by a transverse bar of bone instead of a vertical forehead. The scientists who described these fossils before World War II were perplexed by this curious morphology, to such an extent that although they wrote a very

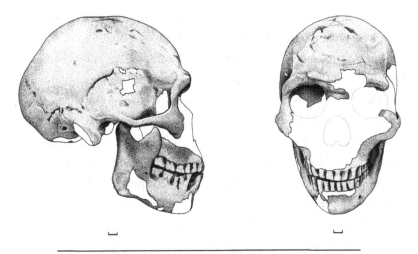

Cranium V from the site of Skhūl, in Israel. Now thought to be well over 100 thousand years old, the Skhūl fossils have for long been assigned to Homo sapiens, *but they are actually quite distinctive in morphology. Drawing by Don McGranaghan.*

large volume on the subject, what they actually concluded, if anything, about the identity of the Skhūl hominids remained spectacularly obscure.

One possibility, of course, was that they represented a hybrid population between moderns and Neanderthals. Geographically this would make sense, for the Skhūl locality is only a few minutes' easy stroll from the cave of Tabūn, long occupied by Neanderthals. Indeed, Neanderthals seem to have been in residence there close to the time that the Skhūl burials were made, although there is no independent reason to believe that they were around at exactly the same moment. Biologically, though, the story is rather different: Neanderthals and moderns were built on a fundamentally different plan; and although we really have no idea at all what a modern/Neanderthal hybrid *should* look like, we do know that hybrids tend to exhibit traits of both parental populations. And this is certainly not what we are seeing at Skhūl.

As a result of this equivocal anatomy, paleoanthropologists have typically yielded in recent years to the temptation to avoid the vexing issue of who these hominids actually were, by brushing them under that "archaic *Homo sapiens*" rug. But this is really evading the issue, and there are other possibilities. One is, of course, that the Qafzeh/Skhūl hominids represent an entirely distinct lineage of hominids of which we otherwise have no record. But a more intriguing possibility is that their curious anatomy was the result not of interbreeding between modern *Homo sapiens* and Neanderthals, but of interbreeding between moderns and the descendants of the North African Aterians exemplified by the Jebel Irhoud and Dar-es-Soltan fossils. We have no idea what we ought to expect from a hybrid between these populations; but somehow this pairing looks like a much more plausible combination, and with pretty recent common African origins the two were presumably very closely related. The prevailing environmental conditions also were right, as was the timing. It is quite possible that, during the wet period around 120 thousand years ago, Aterians spread eastward across North Africa and then turned north to cross a relatively hospitable Sinai Peninsula into the Levant, while at the same time sub-Saharan Africans were moving directly north along the Nile Corridor, dog-legging thence up into Israel. At some point, the two kinds of hominid would have encountered each other, and might have successfully interbred despite their physical differences. Certainly, they were a lot less differentiated from each other than the moderns were from Neanderthals. Precisely how Mousterian methods of stoneworking came to be adopted in the new land remains obscure, though they were not greatly different in concept from either ancestral toolmaking manner, and various North African stone tool assemblages have at one time or another been identified as "Mousterian." As DNA technology improves, perhaps a way will become available of testing the many complex scenarios that potentially arise from this particular possibility of intermixing.

Meanwhile, the presence of a couple of thoroughly anatomically modern *Homo sapiens* at Jebel Qafzeh would, on the face of it, reinforce the notion of a mixed—or rather, newly mixing—population, while at the same time demonstrating a local association between anatomically modern hominids and the Mousterian. Whoever the Qafzeh moderns were, they were not demonstrably behaving in any significantly different manner from Neanderthals. The same thing goes for Skhūl, where the lithic context is also definitely Mousterian. But at Skhūl the picture may be complicated a bit by the recently reported presence both of pigments and of shells that were apparently pierced for stringing. We will come back to this in a moment; meanwhile, what is clear is that the early foray of anatomical moderns outside of Africa, whatever exact form it took, was ultimately unsuccessful. By 60 thousand years ago Neanderthals seem to have been back in charge of the Levant, and we see no more evidence of *Homo sapiens* in the region until much later, by which time our species had contrived to establish the cognitive and technological superiority that it had evidently lacked earlier.

Altogether then, in the light of the frustratingly little we know, it seems reasonable to see the initial excursion of anatomical *Homo sapiens* out of Africa and into the neighboring Levant as the fortuitous product of circumstance, facilitated or even spurred by a benevolent change in climate. Later on, the new people beat a retreat back into their native continent (or more probably died out on the spot), quite likely driven by a climatic deterioration, for we know that conditions became extremely arid during that cold snap about 60 thousand years ago. This drying event severely hit those original Aterian populations in the Sahara, too; and by around 40 thousand years ago this culture was limited to a few lingering outposts along the Mediterranean coast. But whatever the ultimate identity and fate of the Aterians, and despite those intriguing fossils at Qafzeh and those tantalizing hints from Skhūl, we have no evidence that *Homo sapiens* managed to launch a widely successful invasion of Eurasia until much later in time.

THE MOLECULAR EVIDENCE

The notion of an initial ancient foray out of Africa by behaviorally archaic *Homo sapiens,* who later found themselves confined again to their natal continent, fits well with the conclusions of molecular anthropologists. Extensive comparisons of DNA data sets obtained from living human populations from all over the world suggest an origin of our species somewhere in the African continent (most likely somewhere in its eastern or southwestern regions). That founding group subsequently expanded south, north, and west to populate the rest of its home continent, and ultimately Eurasia and the world. With this spread came population expansion and local diversification; and within the African continent at least 14 distinct modern lineages descended from the ancestral population have been identified, each with its own variants. This represents a degree of genetic diversity which, compared to data from the rest of the world, suggests on its own that humans have been evolving in Africa for longer than they have been elsewhere. But to foreclose any argument, all of the major genetic lineages found in other parts of the world are best interpreted as diversified subsets of the variety found in Africa, again pointing toward an African origin for our species. Interestingly, molecular researchers have found that their conclusions are also broadly supported by linguistic and cultural divisions, despite the fact that cultural innovations (because they can be transmitted laterally within the same generation) are subject to weaker constraints than those controlling the spread of biological innovations.

Another set of molecular studies has concluded not only that the founding population was African, but that it was very small. It turns out that, for all the structured DNA variety we find in human populations, this variety is not very impressive when we compare it to what we see in other species, even close relatives. A single population of chimpanzees in West Africa, for example, is said to show more diversity in its mtDNA than the entire human species does today. This can mean one of two things, or both: that our species itself has a recent origin, hence has not had a very long time in which to diversify; or that the founding population was very small. In the event, both of these factors appear to have played a role. *Homo sapiens* seems to have separated from its (now extinct) closest relative only about a tenth as long ago as the two surviving chimpanzee species appear to have split. And, although we don't know what extinct relatives the chimpanzees might have had, it's clear that by general mammalian standards *Homo sapiens* is a very young species. But that's not all. Close analysis of the way in which human DNA variants are distributed today also reveals a pattern strongly suggesting that the ancient human population passed through one or more bottlenecks, or severe contractions, over the course of the late Pleistocene. The most significant of these bottlenecks plausibly occurred around the time at which both archaeological and paleontological indicators imply that people who were both anatomically and intellectually modern first left Africa, ultimately to populate the world.

The dates and duration of the bottleneck vary a bit depending on which of the available data you are looking at; but broadly this event appears to have taken place at some time between 75 thousand and 60 thousand years ago. I include the earlier date mainly because of one scenario that points to a hugely dramatic environmental cause as the main culprit: the explosive eruption of the Indonesian volcano Mount Toba. Around 73.5 thousand years ago, Toba was blasted apart by what was certainly one of the largest and most violent volcanic eruptions in recent geological history. This event devastated the local area, and millions of tons of fine ash were thrown into the air in a cloud that possibly lasted for years, blocking incoming sunlight and causing a "volcanic winter" that affected regions throughout the Old World. It's also been argued that, in combination with the effects of a subsequent drop in world temperatures at the beginning of MIS 4 about 71 thousand years ago, this winterizing effect would have contributed to a dramatic decrease in hominid populations, including that of the nascent *Homo sapiens* in Africa. And while many doubt that Toba's antics, destructive as they doubtless were, would have had effects quite as far-flung as this scenario implies, what is almost certain is that the cold MIS 4 (about 71 to 60 thousand years ago) took its toll on hominid populations everywhere in the Old World.

In Africa the onset of this harsh spell ushered in the extended period of drought that expelled the Aterians from the Sahara, and there's no doubt that it severely afflicted other hominid populations too. As we've seen, the kind of environmental disruption caused by this climatic deterioration is just the kind of thing that promotes response in small, fragmented populations. And it is more than plausible that one local African population of *Homo sapiens,* emerging from this environmental trial as fully symbolic, went on to populate the world. For the first stirrings of the symbolic spirit were already visible well before the stresses of MIS 4 took hold.

To complete the picture of the emergence of the human species and its occupation of the world, molecular anthropologists have been able to draw in the routes by which humans colonized the globe by studying the distributions of various DNA markers in diverse populations. Allowing for the different data sets used (e.g., mtDNA, Y-chromosomes, and various nuclear DNA markers), they have been able to do this with remarkable precision, and to a very fine level of historical detail. The fact that males turn out to have had different migration histories from females complicates things; and although it is actually quite understandable given the typical differences between the human sexes in social and economic roles, it nevertheless confuses the histories of populations as wholes. When it comes to *Homo sapiens,* it seems, nothing is simple. Despite all the complications, though, the various molecular scenarios fit reasonably well with what we know of the fossil record, sketchy as it is.

Apart from those early and unsophisticated émigrés in Israel, we do not have any clear *Homo sapiens* fossils from anywhere outside Africa earlier than the molecular evidence suggests we should find them. A fragmentary mandible some 100 thousand years old from the cave of Zhirendong in southern China has recently been touted as that of a *Homo sapiens;* but its features actually group it most plainly with the endemic "Peking Man" *Homo erectus,* rather than with any potential early modern invaders. Broadly, the molecules indicate that in the period following about 60 thousand years ago, as the rigors of MIS 4 were giving way to the kinder conditions of MIS 3, the bearers of several African DNA lineages left the parent continent. The first principal migration was via Asia Minor into India, whence a coastal route was followed into southeast Asia. All this happened quickly: as we know from archaeological evidence, humans were in Australia by at least 50 thousand years ago. This is all the more remarkable because the first Australians must have crossed at least 50 miles of open ocean to reach their new home: a feat that would have required not only boats—or at least, sophisticated rafts—but excellent navigational skills as well.

Meanwhile, one branch of the migrants continued down into southeast Asia, and another went north to colonize China and Mongolia, eventually doubling back into Central Asia. Migrants of African origin reached Europe, presumably via Asia Minor, by about 40 thousand years ago; and even as climatic conditions were descending to the trough of the last glacial period some 21 thousand years ago, modern people had ventured as far as northern Siberia, above the Arctic Circle. The extraordinary extent of this culturally enabled achievement, involving as it did survival in some of the most difficult conditions the world has to offer, is emphasized by the fact that the supposedly cold-adapted Neanderthals had tended to shun such environments by hundreds of miles.

The process that led to the modern human takeover of the Old World—and later of the New World and the Pacific—was not, of course, one of deliberate expeditioneering. Humans almost certainly expanded their ranges largely by simple demographic spread, as populations grew and budded off new groups into new territories. Of course, as local conditions fluctuated, the process would neither have been regular nor inexorable. Tiny populations were certainly washed back and forth by constant climatic and demographic vicissitudes, with numerous local expansions and extinctions. But this didn't mean that on balance human spread could not occur rapidly: if a human population expanded its range by only ten miles in a generation, this would add up to more than 1,500 miles in a mere 2,500 years, something quite feasible on the timescale involved. Whatever the details, though, just in itself population growth on this scale implies that there was something *different* about these new migrants: namely, an unprecedented ability to intensify their exploitation of the environments around them. This fed back into growing populations, and consequently into further geographical expansion.

This demographic difference is also implicit in the fact that the new *Homo sapiens* was not moving into territory that was virgin for hominids: related species almost certainly already resided in many if not most of the regions into which it expanded. And the larger pattern their encounters took is clear. When behaviorally modern humans moved into Europe, the behaviorally archaic Neanderthals yielded. When they moved into southern Asia it was *Homo erectus,* which flourished equally late in its last southeast Asian island redoubt, that promptly disappeared. The same went, a little later in time, for those unfortunate Hobbits of Flores; and probably also in poorly documented Africa for any other hominids that may have survived the rigors of MIS 4. There was clearly something *special* about the new invaders. From the very beginning of hominid history, the world had typically supported several different kinds of hominid at one time—sometimes several of them on the very same landscape. In striking contrast, once behaviorally modern humans had emerged from Africa the world rapidly became a hominid monoculture. This is surely telling us something very important about ourselves: thoughtlessly or otherwise, we are not only entirely intolerant of competition, but uniquely equipped to express and impose that intolerance. It's something we might do well to bear in mind as we continue energetically persecuting our closest surviving relatives into extinction.

THIRTEEN

THE ORIGIN OF SYMBOLIC BEHAVIOR

Our ancestors made an almost unimaginable transition from a nonsymbolic, nonlinguistic way of processing and communicating information about the world to the symbolic and linguistic condition we enjoy today. It is a qualitative leap in cognitive state unparalleled in history. Indeed, as I've said, the only reason we have for believing that such a leap *could* ever have been made, is that it *was* made. And it seems to have been made well *after* the acquisition by our species of its distinctive modern biological form.

The earliest firm intimations we have that a symbolic sensibility was astir among populations of newly evolved *Homo sapiens* come from Africa or its immediate environs. The oldest of them are also a bit arguable, consisting principally of suggestions that at Skhūl, over 100 thousand years ago, small marine snail shells were already being pierced for stringing as beads, while lumps of pigment were heated, presumably to change their color from yellow into a more attractive orange or red. The shell beads are particularly interesting, because personal ornamentation using necklaces or bracelets (and, for that matter, bodily coloration using pigments) has usually had deep symbolic significance among historically documented peoples. How you dress and decorate yourself signifies your identity as a member of a group, or of a class or a profession or of an age-cohort within your group. Still, the early putative evidence for this kind of thing is slender at this point: two shells perforated (possibly by natural causes) through their weakest points at Skhūl, and a single shell at an Aterian site in Algeria of uncertain age. At both places, however, the shells were of a species that had to have been collected far away along the Mediterranean shore. This implies they were special objects to their possessors, specifically brought in through long-distance exchange. Possibly yet more significantly, at both sites the beads were made from the shells of a genus, *Nassarius*, that was widely used later on elsewhere in better-substantiated ornamental contexts.

Somewhat firmer bead evidence starts turning up at around the 80-thousand-year point at other Aterian sites. Thus a dozen perforated *Nassarius* shells were found at the Moroccan Grotte des Pigeons. As at the Israeli and Algerian sites, this cave is far from the nearest possible source of the shells, which suggests that they had been deliberately brought in, probably through trade. But, again, there is as yet no definitive evidence that the holes in them were produced by human agency— though a few shells bore traces of pigment suggesting that they might have been deliberately colored, and showed a curious polish that may have come from rubbing against someone's skin.

Similar finds at other sites along the North African coast suggest that the Grotte des Pigeons findings were not isolated. But the best evidence suggesting that these North African manifestations were part of a larger pattern comes, oddly enough, from the opposite end of the African continent, some four thousand miles away. At Blombos Cave, a coastal site not far from the continent's southern tip, archaeologists found numerous perforated *Nassarius* shells that were worn in a way that strongly implies that they had been strung as beads. With a Middle Stone Age industrial context, and dating to about 76 thousand years ago, these beads were convincingly in the same general time range and broad cultural context as their North African equivalents, and their general acceptance as objects of personal adornment has made it look probable that some African MSA populations, at least, had begun to decorate their bodies in the period following about 100 thousand years ago.

But these people also left evidence of more explicitly symbolic behaviors. For Blombos has given us the earliest objects that we can confidently interpret as symbolic. These come from the same Middle Stone Age levels that yielded the *Nassarius* beads, and they consist of a couple of ochre plaques several inches long, on each of which a deliberately smoothed surface displays a distinctly engraved cross-hatched pattern. What the pattern was intended to convey we may never know; but the two pieces were found several vertical inches apart in the deposits, suggesting that this geometric pattern was a consistent one that retained its meaning over time, and didn't simply represent a single individual's idle doodlings. Another piece of ochre, found at a rock shelter some 250 miles away and possibly created only fractionally later in time, bears what may be a simplified version of the same motif, providing yet more evidence that the pattern contained meaning. What's more, in the same deposits at Blombos were found bone tools, probably originally hafted, of the kind that are so conspicuously missing from the Neanderthals' contemporary tool kits in Europe.

Not too far from Blombos is another coastal cave complex, at Pinnacle Point. This was also inhabited by MSA humans, beginning around 164 thousand years ago and continuing with hiatuses (possibly because high sea levels washed out intervening deposits during MIS 5e) until well under 70 thousand years ago. At the 164-thousand-year point, occupants of these caves were already expanding their diet to incorporate hard-to-obtain marine resources, possibly in response to the cold climatic conditions of MIS 6. At the same time, they were regularly processing

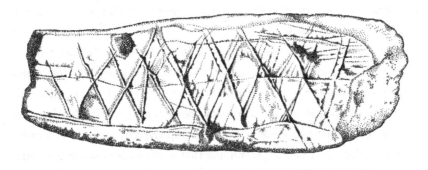

One of the geometrically engraved ochre plaques from Blombos Cave on the southern African coast. Around 77,000 years old, this piece is the earliest evidently symbolic object. Drawing by Patricia Wynne.

pigments and producing "bladelets"—small stone flakes that were sunk into handles, the likes of which were not seen outside of Africa until very much later. This was clearly a time of significant cultural change, though it is arguable that any of these technological expressions was *necessarily* a harbinger of symbolic cognition.

Still, while I've already claimed that there is very little in Old Stone Age technology that we can take as *prima facie* evidence of the workings of symbolic minds, if there is an exception to this rule we find it at Pinnacle Point. Good materials for stoneworking are pretty rare in the area; and there is good evidence that by around 72 thousand years ago the people who stayed there were using complex technology to improve at least one of the indifferent raw materials available to them: silcrete, a type of rock that occasionally forms in silica-rich soils. Silcrete is suitable for flaking by stone tool makers, but in its natural state it doesn't hold an edge for long. However, the Pinnacle Point people discovered that, if it is appropriately heated and cooled in an elaborate series of steps, silcrete hardens and makes much better tools. The technology involved is so complex, and involves so many stages of forward planning, that it could almost certainly never have been imagined and carried out by minds incapable of abstracting and visualizing long chains of cause and effect. So at 72 thousand years ago, right around the time when the Blombos people produced those symbolic plaques (but in an entirely different way), their Pinnacle Point neighbors were also exhibiting dawning symptoms of symbolic reasoning. Silcrete tools occur in the earlier levels of the Pinnacle Point complex as well, but at that more remote time the evidence is less good that the material from which they were made was deliberately heat-treated.

As if in riposte to the findings at Pinnacle Point, the latest report from Blombos indicates that, around 75 thousand years ago, the people there were also exhibiting advanced technological prowess. For not only have fire-hardened silcrete tools now been confirmed at Blombos, but it's recognized that their edges had been improved by an approach called pressure flaking: a delicate technique that is otherwise known only after about 20 thousand years ago in Europe, where it was the work of late Cro-Magnons. This unexpected finding is pretty amazing, but it does make it much easier to believe a reported date of around 90 thousand years for some bone harpoons found at a site called Katanda in the Eastern Congo. This is many tens of thousands years than the earliest previously known barbed harpoons, made by Cro-Magnons in Europe well after the appearance there of pressure flaking. Evidence is thus beginning to pile up very impressively that something really important was astir in Africa during the middle and later phases of the MSA.

Exactly who the southern MSA people were remains a conundrum. They do not seem to have practiced burial at or near their living places, so few of their bones have been preserved. Several isolated teeth from Pinnacle Point are not very informative, but one notable exception to the lack of human fossils in association with the MSA occurs at the caves of the Klasies River Mouth, a couple hundred miles east along the coast from Blombos. In MSA levels at Klasies, archaeologists found evidence for a symbolic division of the living space into distinct functional areas by over 100 thousand years ago; and at 80 to 90 thousand years ago, some highly fragmentary human remains were discovered that may represent the cooked remnants of a cannibal feast. The remains themselves have usually been interpreted as those of *Homo sapiens*; and if they do not fall precisely within the envelope of variation shown by our species today, they are certainly very close.

Following all the excitement of Blombos times, the symbolic trail cools a bit. A few MSA sites in southern and eastern Africa have produced fragments of ostrich eggshell bearing what seem to be deliberately incised patterns. Such pieces are most numerous at South Africa's Diepkloof rock shelter. At 60 thousand years or so, this site is a bit later than Blombos, but it is still firmly in the MSA. The fragments seem to have originally formed part of water containers decorated with symbolic motifs, and they convincingly confirm an ongoing South African symbolic tradition during the later part of the MSA. Ostrich eggshell was also used as a material for symbolic artifacts at East African sites, though later in time. The best-known such locality is Enkapune Ya Muto (EKP) rock shelter, in the Kenyan Rift Valley. EKP is dated to about 40 thousand years ago, which places it at the local point of transition from the MSA to the Late Stone Age (LSA). The LSA is a period of African prehistory that originated at around the same time the Upper Paleolithic arrived in Europe, and the two are considered broadly equivalent; as far as we can tell, the LSA was the work everywhere of fully modern humans. The EKP ostrich shell objects are unengraved but beautifully shaped disks that were strung as beads; so they differ in category from the decorated containers from Diepkloof. Nonetheless, some cultural continuity with the South African LSA may be hinted at by similar beads found at South African LSA sites.

Africa is a very large continent, vast tracts of which are *terra incognita* to Paleolithic archaeologists. It is thus particularly frustrating that, apart from such tantalizing finds such as those from Katanda, we really don't have much of a clue as to what was happening midcontinent in the symbolic domain, at the time when hominids both in the far north and the far south were showing those intriguing signs of dawning modernity in their ways of dealing with the world. How much indirect contact there may have been between human populations in the north and south of Africa during the later MSA has to remain largely conjectural, as does just how much those populations may have been differentiated biologically from each other. The fact that *Nassarius* shells were the chosen bead-making objects early on in both regions is suggestive, but perhaps not a lot more than that. During the wet phase 120 thousand years ago the dramatic ecological barrier that nowadays separates the northern fringes of the African continent from its sub-Saharan south really didn't exist, and it is not unreasonable to suppose that in a period of awakening symbolic awareness, regular trading routes crossed what has now become the hyperarid Sahara. But during the following MIS 4 the desert reasserted itself as a formidable barrier, at least intermittently; and by the time that we find those intriguing hints of a new cognitive style around 80 thousand years ago, it is by no means clear that routine cultural and biological interchange would have been possible, except maybe via the Nile Valley.

Still, if the molecular anthropologists are right (and they are backed up by the archaeological record, as far as it goes), the cognitively modern humans who eventually took over the world originated in an eastern African population, probably a rather small one, that very likely lived *after* Blombos times. It is also probably relevant in this respect that, not long after the spurt of symbolic innovation between 80 and 60 thousand years ago, southern Africa was plunged into a prolonged period of drought during which its interior became largely depopulated. Under these conditions it is surely only moderately likely, at best, that the cultural practices of the Blombos and Pinnacle Point people were directly antecedent to later expressions of similar kind farther to the north. It seems far more likely that there was a quickening of the creative hominid spirit in a variety of places during the later part of the MSA, capitalizing everywhere on a general predisposition toward symbolic thinking that had been born rather earlier, with the origin of *Homo sapiens* as a hugely distinctive physical entity at some time after the MSA had begun.

Wherever in Africa the population ancestral to all living humans emerged, and whatever route its descendants took during their exodus, it is clear that cognitively modern humans had made their appearance in Eurasia by not long after 60 thousand years ago, at latest. We've seen that humans were in Australia by around 50 thousand years ago; and they were leaving traces of artistic activity there not long thereafter. Sites in southern India that may also be in this date range yield stone tools remarkably similar to those that were produced at sites in southern and eastern Africa at places such as Blombos, Diepkloof, and EKP; and at one of them fragments of ostrich eggshell have been found that bear cross-hatched motifs similar to those from Blombos and Diepkloof. Most remarkably of all, we have evidence both from the physical remains and the art of the Cro-Magnons that people with an entirely full-blown modern sensibility had arrived in Europe, a relatively remote and inaccessible peninsula, soon after we begin to find shell beads at early Upper Paleolithic sites in Lebanon and Turkey. This evidence dates to over 40 thousand years ago and confirms that Cro-Magnon predecessors were already spreading north and west by that time.

The Cro-Magnons have left us all the evidence we could possibly need that they were people cognitively like us. Yet the fact that we have this proof of the Cro-Magnons' cognitive attainments is largely an accident of both cultural expression and topography. Decorating the dank and dangerous depths of caves with fabulous animal images and a whole vocabulary of geometrical symbols is, to put it mildly, a rather unusual pursuit; and while all historically documented human societies have clearly been symbolic, few others have expressed this capacity in so durable a way. What's more, most Cro-Magnon art was preserved in caves and crannies in the limestone landscapes these people just happened to inhabit; in other geological settings we might not expect parallel expressions to be preserved. Still, the accident is a happy one indeed, and it certainly gives us a minimum date for the attainment of fully modern consciousness.

Some scholars have suggested that the dazzling Cro-Magnon art represented such a break with the past that a recent genetic modification must have been acquired in the Cro-Magnons' lineage to make all this creativity possible: a modification whose effects were confined to their neural information processing, and were not reflected in the fossil bones which are all the physical evidence we have of them. But biologically this seems less plausible than other possibilities; and the hints from Blombos and Pinnacle Point give us excellent reason to believe that the symbolic sensibility, of which Cro-Magnon art was the finest and most exhaustive early expression, was already initially astir much earlier in human history. Beyond this, there really is no reason whatever to think that the earliest possessors of the cognitive capacity that underwrites this sensibility should have discovered all of its dimensions at once. Indeed, the subsequent technological and economic histories of humankind have been virtually synonymous with the exploration of this relatively new-found capacity. And we are still groping toward its invisible limits today.

FOURTEEN

IN THE BEGINNING
WAS THE WORD

I n biological terms the birth of anatomical *Homo sapiens,* maybe as much as a hundred thousand years before the first *Nassarius* bead was ever pierced, was a huge event. We differ from our closest known relatives in numerous features of the skull and of the postcranial skeleton, in important features of brain growth, and almost certainly in critical features of internal brain organization as well. These differences exist on an unusual scale. At least to the human eye, most primate species don't differ very much from their closest relatives. Differences tend to be largely in external features such as coat color, or ear size, or even just in vocalizations; and variations in bony structure tend to be minor. In contrast, and even allowing for the poor record we have of our close extinct kin, *Homo sapiens* appears as distinctive and unprecedented. Still, we evidently came by our unusual anatomical structure and capacities very recently: there is certainly no evidence to support the notion that we gradually became who we inherently are over an extended period, in either the physical or the intellectual sense. As I have already observed, this suggests that the physical origin of our species lay in a short-term event of major developmental reorganization, even if that event was likely driven by a rather minor structural innovation at the DNA level. Such an occurrence is made more plausible by the fact that genetic innovations of the kind that probably produced us are most likely to become "fixed" (i.e., the norm) in small and genetically isolated populations, such as those into which climatic vagaries would regularly have fragmented our already thinly spread African forebears. In other words, conditions in the late Pleistocene would have been as propitious as you could imagine for the kind of event that would necessarily have had to underwrite the appearance of a creature as unusual as ourselves.

As far as we know, *Homo sapiens* is totally unique in significantly expressing an ability to manipulate information symbolically. And understanding how we acquired this capacity is fundamental to any complete understanding of ourselves. Some possibilities we can eliminate right away. To begin with, our novel way of dealing with information was hardly a predictable outcome of any identifiable trend that preceded it. And neither was it simply a threshold effect of acquiring a greater and greater brain volume over vast spans of time, as smarter individuals outreproduced dumber ones in our ancestral lineage. We know this not only because the nonsymbolic Neanderthals had brains that were on average larger than ours, but because our own brains appear to have shrunk by as much as ten percent since Cro-Magnon times, and we haven't slipped below the symbolic threshold yet. Whatever you want to make of this latter tidbit, it is at the very least evident that we have to look to more than simple increase in brain mass to explain our unusual cognitive style.

The only evident alternative is that our strange intellectual faculty is attributable to a novel neural conformation, a change in the internal organization and wiring of our brains. Acquisition of such a novelty would not in itself be unprecedented; after all, the human brain has a long and largely accretionary history going right back to the earliest vertebrate brains half a billion years ago—and beyond. Nothing inherently new there. But the *results* of this acquisition were revolutionary: in today's jargon, they were "emergent," whereby an adventitious change or addition to a pre-existing structure led to a whole new level of complexity in function.

Exactly when our amazing capability was initially acquired is something we cannot read directly from the fossil record: the paleoneurologists, those specialists who specialize in the form of fossil brains as determined from the impressions they leave inside the cranial vault, cannot even agree in principle if there is any functional significance to the minor external shape differences we see between modern human and Neanderthal brains. All we know for sure, from the archaeological evidence, is that the two species *behaved* differently. The Neanderthals seem to have possessed a sophisticated version of the "old-style" hominid way of dealing with stimuli, using purely intuitive processes. In contrast, we symbolic *Homo sapiens* are processing information in an entirely revolutionary and unprecedented way, even though the "old" brain is still very much there, deep inside.

GENES, LANGUAGE, AND LARYNXES

I mentioned briefly that some believe our new way of doing business is due to the very recent acquisition of a "symbolic" gene. Excitement about this possibility ran high after scientists discovered that the human version of a gene called FOXP2 was intimately involved in our language abilities, at least to the extent that people possessing a mutated form are unable to speak properly (though they don't seem to be more broadly cognitively impaired). Imaging studies showed that such people have reduced activity in Broca's area of the brain. The excitement increased yet further when various Neanderthals were shown to have possessed the normal human version of the FOXP2 gene, sparking speculation that here was proof that the Neanderthals had possessed language. If this speculation had been on the mark, it would have been huge in its implications for complex consciousness; for language is a supremely symbolic system that depends on the creation and manipulation of mental symbols for its very existence. Any organism able to generate language would almost certainly have been capable of exhibiting all of the other correlates of symbolic thought. But if only it were that simple: it turns out that numerous genes (all of them working as they should) are involved in determining normal language and speech production in humans. Indeed, so many genes are intimately choreographed in the developmental process that sometimes it seems a miracle that any of us ever develops normally. In view of all this it is evident that the notion of a single gene "for" language (even a regulating gene, like FOXP2) is an illusion, albeit an attractive one. What the Neanderthals possessed was a necessary condition for language, but it wasn't sufficient.

So far at least, then, there are no "silver bullet" genes that we can finger as the root cause of our cognitive uniqueness. But it turns out that there is a much better general explanation for our possession of a brain anatomy that facilitates complex thought: one that, moreover, fits far better with the little we know of the behavioral record at the time this uniqueness seems to have begun expressing itself. The specifics still evade us, and we have as yet no idea what the genetic rearrangement was that gave rise to the unique anatomy of *Homo sapiens*. All we know for sure is that this event did indeed occur. But it seems overwhelmingly likely that—like all of our other unique attributes of structure—our new cognitive ability was acquired as a byproduct of the hugely ramifying genetic accident that resulted in the appearance of *Homo sapiens* as a distinctive entity. Happily for us, the resulting creature turned out to function pretty well.

In this view, the addition of the neural ingredient that predisposed our species for symbolic thought was simply one passive consequence of the developmental reorganization that gave rise to anatomically recognizable *Homo sapiens* some 200 thousand years ago. And it seems justifiable to look upon what happened as analogous to the construction of an arch, which cannot function until the keystone has been dropped into place. What's more, whatever the "keystone" was in our case, the new potential it created then lay fallow for a substantial length of time, until its symbolic potential was "discovered" by its owner.

Although it may seem a little counterintuitive, this time lag between the acquisition of what turns out to be a very significant novelty and its exploitation by its possessor, is actually an example of a very common phenomenon in the evolutionary history of life. Since all genetic innovations occur at random relative to the circumstances of their carriers' existences (though they may be channeled, of course, by their owners' evolutionary histories), they must arise initially not as *ad*aptations to a particular lifestyle, but as *ex*aptations: features that must necessarily be co-opted *post hoc* into a new use. I've already briefly mentioned the classic example of feathers, which were possessed by the ancestors of birds many millions of years before these modified dermal follicles were ever

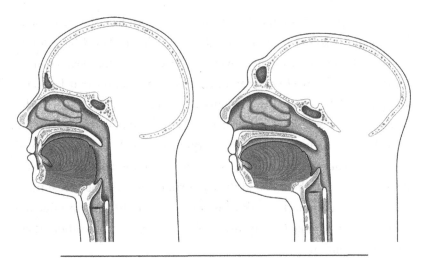

Cross-sections through the heads of a modern human (left) and a Neanderthal (reconstructed; right), to show differences of the upper vocal tract. Note the long palate and tongue of the Neanderthal, and the higher placement of its larynx compared to the Homo sapiens. *Illustration by Diana Salles, after sketches by Jeff Laitman.*

recruited as essential components of the flight mechanism. Similarly, the ancestors of terrestrial vertebrates had already acquired the rudiments of legs while they were still fully aquatic, and a terrestrial existence was still far in their future. You simply wouldn't have predicted their future function when they first appeared. What is more, evolutionary novelties often persist if they don't actively get in the way; and in the case of *Homo sapiens* the potential for symbolic thought evidently just lurked there, undetected, until it was "released" by a stimulus that must necessarily have been a cultural one—the biology, after all, was already in place.

This biology included not only the cerebral potential for generating language and passing instructions for its production along to the peripheral vocal structures, but also those peripheral structures themselves. There has been a lot of argument over just what it is about our upper vocal tract that allows us to produce speech, and without which our audible language could not be expressed. Much of this discussion has involved the low position in the human throat of our larynx (voice box), and how you might recognize in fossils exactly where the larynx lay. The lower the larynx is, the more pharynx (airway) there is above it that can be manipulated, by the throat muscles, to produce the frequencies that emerge as sound from the vibrating air column. Many have found reason to believe that larynxes were lowered to varying degrees in an assortment of fossil *Homo* crania, sparking speculation that language abilities (and by extension, human-style consciousness) began forming early in the evolution of our genus. Yet even the discovery of fossil hyoid bones (the bony part of the larynx) has done nothing much to quell argument over this matter, and lately attention has shifted toward the proportions of the oral and throat portions of the upper vocal tract, and to the suggestion that a short face is necessary to allow production of the necessary range of frequencies. We can expect debate about these features to continue.

Meanwhile, though, there is a very attractive feature of the notion that the potentials for language, for speech and for symbolic thought were instead born together, at the origin of anatomical *Homo sapiens*. This is, that *all* of the necessary features would already have been in place by the times that they were co-opted (independently) for their new uses. Among other things, whatever functional context the short, retracted face of *Homo sapiens* had evolved in, it had nothing to do with language or perhaps even speech. It is hard to tell just what the actual context might have been, especially since a small, retracted face comes with some significant disadvantages. For one, it reduces the length of the tooth-rows, crowding the teeth and frequently leading to impactions and malocclusions; for another, lowering the larynx involves a crossing of the airways with the food tract that introduces a severe danger of choking—something that is much less likely when the larynx is high. This effect is more than an inconvenience; in Japan alone, famous for its bite-sized foods, more than 4,000 people choke to death every year. It's anybody's guess what the countervailing early advantages of the new skull form might have been. Maybe the disadvantages just weren't enough to make a significant difference, or maybe the slender new body build was energetically economical enough to provide a competitive advantage with more massively constructed and lower-mileage competing hominids. But clearly, what proved up to the job was the new and unusual human organism as a whole, rather than specific aspects of its innovative anatomy.

Still, early *Homo sapiens* did not overwhelm the competition right away. As we have seen, its initial and apparently unsymbolic foray into the Levant was not a permanent success. Instead, the rapid takeover of the world by our forebears had to await the arrival of symbolic behavior patterns. The spotty evidence we have of mankind's symbolic awakening does not rule out either of two possibilities as to just how this development happened within the African continent. Given that the biological potential was already present, multiple isolated hominid populations in various parts of Africa might have started experimenting with the new ability; or there was just a single point source. Knowing for sure which was the case will require a lot more information than we have at present, though the broad distribution of early rumblings suggests that, at the very least, symbolic information processing was an idea whose time had arrived by 80 thousand years ago.

SYMBOLIC AWAKENINGS

Exactly how the almost unimaginable transition to the symbolic mental manipulation of information took place remains a subject of pure speculation, though an irresistible one. We have already established that we need to look for a cultural stimulus that kicked the biologically pre-enabled human brain into symbolic mode. If you asked an assortment of scientists interested in this question what that stimulus might have been, two clear frontrunners would probably emerge.

One of these potential stimuli is "theory of mind." We humans are primates, and our higher primate relatives in general are intensely sociable. Yet we display a particular kind of sociality, characterized not only by the kind of prosociality—concern for others—that the apes don't seem to share, but also by a more detached, observational sociality. We know what we are thinking (known to psychologists as "first-order intentionality"), we can guess what others are thinking (second-order), we can suspect that someone else has a belief about a third party (third-order), and so on. Apes seem to have achieved first-order intentionality, and alone among nonhuman primates may have clambered on to the second level; humans, on the other hand, seem to be able to cope with up to six levels of intentionality before their heads begin to spin (he believes that she thinks that they intend . . . and so on). Some scientists believe that that the evolution of our extraordinary cognitive style was driven by the development of the increasingly elaborate theory of mind needed to cope with the dynamics of interaction within societies that were steadily becoming more complex. In other words, modern human cognition developed under the self-reinforcing pressures of increasingly intense sociality—maybe around those campfires.

This is an attractive idea, especially as our elaborate social rituals and responses are so intimately interwoven with our ways of processing information about our fellow members of society—always a subject of intense preoccupation to us. But a mechanism of this kind explains neither why the highly social apes haven't developed a more complex theory of mind over the time during which they have been evolving in parallel with us, nor why the archaeological record seems to indicate a very late and essentially unheralded arrival of symbolic consciousness in just one lineage of large-brained hominid.

The other thing everybody associates with our cognitive style is our use of language. Indeed, it is hardly overstating the case to characterize language as the ultimate symbolic activity, allowing as it does the generation of an infinite number of statements from a finite group of elements. Like thought, language involves dissecting the world around us into a huge vocabulary of symbols that are then combined, according to rules, to make statements not only about the world as it is directly perceived, but also as it might be. And it is virtually impossible to imagine our thought processes in its absence, for without the mediation of language those thought processes would be entirely intuitive and nondeclarative, merely involving the association of incoming stimuli with remembered ones, and responding accordingly. This is not to say that responses of that kind need necessarily be simple. Extremely complex associations may be made without requiring the process of abstraction that lies at the basis of symbolic thought. We know this from the example of earlier hominids. These precursors did not just get by on this level of functioning, but made some of the most notable technological advances in hominid history, including the domestication of fire, the invention of compound tools, and the building of shelters. Such achievements are impressive indeed. But language facilitated the imposition of symbolic information processing upon older cognitive processes. And this added an entirely new dimension to the way in which hominids saw the world, and eventually reimagined it.

That this momentous event took place in Africa—the continent in which we find the first fossil evidence of creatures who looked just like us, and (somewhat later) the earliest archaeological suggestions of symbolic activities—is corroborated by a recent study of the sounds used in spoken languages around the world. The study of comparative linguistics makes it clear that languages have evolved much as organisms have done, with descendant versions branching away from the ancestral forms while still retaining for some time the imprint of their common origins. Many scientists have accordingly used the differentiation of languages as a guide to the spread of mankind across the globe. And in doing this they have traditionally concentrated on the words that make up those languages. But this has proved a tricky endeavor, for individual words change quite rapidly over time: so rapidly that beyond a time depth of about five thousand years, or ten at the very most, it turns out to be fairly hopeless to look for substantial traces of relationship. As a result, while language has indeed proven useful in tracing the movement of peoples around the Earth over the last few thousand years, linguists have been somewhat stymied when it comes to its very early evolution.

The New Zealand cognitive psychologist Quentin Atkinson has recently suggested an alternative. According to Atkinson, in seeking the origins of language we are better off looking not at words as a whole, but at the individual sound components—the phonemes—of which they are comprised. This makes sense, because the phonemes are much more bound by biology than are the ideas that their combinations represent. And when Atkinson looked at the distribution of phonemes in languages around the world, he found a remarkable pattern. The farther away from Africa you go, the fewer phonemes are typically used in producing words. Some of the very ancient "click" languages of Africa, spoken by people with very deep genetic roots, have over a hundred phonemes. Our English language has about 45; and in Hawaii, one of the last places on Earth to be colonized by people, there are only 13. Atkinson attributes this pattern to what is known as "serial founder effect": a phenomenon, well known to population geneticists, that is due to the drop in effective population size each time a descendant group buds off and spreads away from an ancestral one. With each successive budding, genetic—and apparently also phonemic—diversity diminishes, because of the bottleneck effect discussed earlier.

The signal of this effect in the five hundred or so languages analyzed by Atkinson is weaker than the one found in the genes, but this difference is plausibly due to the rapidity with which languages evolve. The key thing, though, is that the genetic and phonemic patterns are essentially the same, and that both point to an origin in Africa. Atkinson's analysis suggests that the convergence point may be in southwestern Africa, which is also in line with one recent genetic study. And his results imply not only that modern *Homo sapiens* originated in a single place, but also that the same thing was true for language (or at least, for the form of language that survives today). In which case, there is a strong argument for a fundamental synergy between biology and language in the rapid takeover of the world by articulate modern people.

THE TRANSITION

There are many reasons why the invention of language is the obvious candidate for the stimulus that tipped our ancestors over the symbolic edge. Although all modern societies are already linguistic, and have been for a long time, we do know from direct observation that structured pidgin languages that substitute signing for sounds can be created quickly, and without external prompting. The most famous example is a signed language spontaneously developed by deaf children in Nicaragua during the 1980s. When the first schools for the deaf were established in that country during the 1970s, they brought together children who had formerly been isolated at home, among speaking relatives who did not sign. Forming a deaf community for the first time, the kids rapidly and independently created a signed language of their own: one that quickly developed many of the complexities of spoken language, though it bore no relation to the Spanish that was spoken around them.

What's more, in an incredible story, an adult human has been observed going through the process of acquiring language, a procedure that clearly involved a dawning recognition that objects can have names, the most basic of symbols. In her book *A Man Without Words*, the sign-language expert Susan Schaller affectingly describes how she realized that a deaf student in her class not only was unable to sign, but was unaware that other people used names to denote objects. This man, whom she called Ildefonso, had been brought up in a hearing household, in isolation from any stimuli that could have helped him grasp that objects had names. What's more, he lacked access to any kind of special education that might have taught him to mentally create and recognize signs. Yet, although withdrawn, he functioned well enough to find his way into Schaller's classroom, and once there, he immediately gave her the impression of being both intelligent and intellectually curious. As she tells it, Schaller initially tried to teach Ildefonso the rudiments of American Sign Language (ASL), but soon perceived that he did not grasp even the concept of signs. Modifying her approach, she eventually achieved a breakthrough. Ildefonso, in a flash of insight, understood that everything had a name. "Suddenly he sat up, straight and rigid. . . . The whites of his eyes expanded, as if in terror. . . . He broke through. . . . He had entered the universe of humanity, discovered the communion of minds." This changed everything about his perception of the world, and, once he had recovered from the emotional flood his dawning comprehension unleashed, he became hungry for signs, demanding new words.

Understandably, after 27 years without language, without symbols, this realization was also hugely traumatic. Schaller writes movingly of Ildefonso's sense of grief as he perceived "the prison where he had existed alone, shut out of the human race." Still, despite the fact that he subsequently had all of the language-learning difficulties and discouragements that every adult experiences, and much more to boot, he eventually learned to converse in ASL.

Schaller's experience with Ildefonso is as close as anyone in our time will ever come to witnessing what the birth of symbolic humanity must have been like, as an adventitiously prepared brain suddenly discovered what it was capable of. Schaller even believes that Ildefonso's condition is more common than one might imagine, and that many people's "intelligent, sane, yet languageless" state may be routinely mistaken by hearing or signing people for simple deafness with failure to sign. If this is indeed the case, maybe we have in Ildefonso at least a sideways glimpse into the prelinguistic human condition—albeit greatly modified by the loss of earlier systems of communication, and of the cognitive sequelae that went along with them.

Unfortunately, Ildefonso was not of much help in determining what would it be like to be a normal *Homo sapiens* with everything except linguistic function. It eventually came to light that he had belonged to a small community of deaf and languageless individuals who communicated by miming, rather than by signing. Instead of describing their experiences concisely by stringing words together according to rules, they acted those experiences out, rather like guests at a dinner party playing charades. This was a hugely cumbersome way of communicating, so much so that once he had grasped the idea of language and had begun to compile an extensive vocabulary of signs, Ildefonso no longer had the patience to use it, and stopped spending time with his former associates. What's more, he proved extremely reluctant to describe his inner life before he acquired language. There was perhaps no way in which he could have explained the difference; and in any event, he just didn't want to publicly relive it. How and to what extent language is separate or different from what we experience as thought thus remains imponderable on the basis of this particular individual's experience.

This is a pity, because knowing the exact difference between mental processes with and without words is critical to understanding the cognitive difference between nonsymbolic and symbolic *Homo sapiens*. Those early nonsymbolic Levantine *Homo sapiens* got along perfectly well, living lifestyles that were much like those of the crafty and accomplished Neanderthals. And while their nonlinguistic state was limiting compared to our own, presumably they and their predecessors were not living in the oppressive cognitive darkness from which Ildefonso was so relieved to have escaped. They were perfectly fine, with lifestyles of a complexity that no organism before them had ever contrived to achieve.

Perhaps we can catch an indirect glimpse of what being a prelinguistic *Homo sapiens* might have been like in the experience of Jill Bolte Taylor, a neuroanatomist who suffered a massive stroke that deprived her for several years of her linguistic capacity. At the age of 37 she lost her command of language, and as a consequence all her memories disappeared and she found herself able to live only in the present. On the other hand, she also felt a sense of peace, and of unaccustomed connectedness to the world around her. Her previous command of language had, it seemed, not only allowed her but compelled her to distance herself from her surroundings; and this, of course, is the essence of the human symbolic facility, which confers the capacity to objectify oneself and remain apart from one's universe.

Taylor's experience, recounted after a full recovery, is a fascinating one. It's nonetheless obvious that a cerebral accident in a speaking adult cannot precisely recreate the functioning of a normal prelinguistic human brain. But there may be another avenue by which we can envisage the nonlinguistic human state. Some psychologists have argued persuasively that young children who have not yet mastered their parents' language do not think, at least in the way that adults do. And it's possible that their mental manipulations of information may resemble those of prelinguistic *Homo sapiens* in some respects. Still, while it's clear that children who aren't yet talking don't think like linguistic adults, their brains are of course immature (especially in the all-important prefrontal cortex, which matures notably late), and they can't make all the connections between different kinds of information that adults do. What's more, they cannot tell us articulately about their mental states, which they can only express through emotional acting out. And this brings us right back to the dilemma of understanding the chimpanzee that we encountered in the first paragraphs of this book.

It is evident from the archaeological record that complex lifestyles, intuitive understanding, and mental clarity are all possible in hominids lacking language in its modern form. In the appropriate context, to be wordless is not to be dysfunctional. Nonetheless, words are a crucial enabling factor in complex cognition. The ability to manipulate words clearly expands and liberates the mind. The more words you have, the more complex a world you are able to visualize; and, on the other side of the coin, when you run out of words you run out of explicit concepts. Nevertheless, given that our language abilities seem to have been somehow grafted on to the earlier cognitive substrate possessed by the first anatomical *Homo sapiens,* our mental lives today are a constant tightrope-walk between the symbolic and the intuitive. Our symbolic abilities explain our possession of reason, while intuition, which is itself probably a curious amalgam of the rational and the emotional, accounts for our creativity. It is the fortuitous combination of the two that makes us the unstoppable if imperfect force in Nature that we are.

The changeover of *Homo sapiens* from a nonsymbolic, nonlinguistic species to a symbolic, linguistic one is the most mind-boggling cognitive transformation that has ever happened to any organism. The details of this transition will probably forever evade us, and almost any scenario we might dream up risks trivializing it. But, with the examples of the Nicaraguan schoolchildren and Ildefonso in mind, perhaps it is not too hard to envision, at least in principle, how language might have emerged in a small community of biologically prepared early *Homo sapiens* somewhere in Africa. Indeed, I am greatly entertained by the notion that the first language was the invention of children, who are typically much more receptive to new ideas than adults are. They always have their own methods of doing things, and they communicate in ways that sometimes deliberately mystify their parents. For reasons that had nothing to do with language, the children concerned already had all of the peripheral anatomical equipment necessary to produce the full range of sounds demanded by modern languages. They must also have possessed both the biological substrate necessary to make the intellectual abstractions involved, and the innate urge to communicate in a complex manner. And almost certainly, they belonged to a society that already possessed an elaborate system of interindividual communication: one that employed vocalization as well as gesture and body language. After all, as in the case of every behavioral innovation, the necessary physical springboard had to have been there. And with the Nicaraguan example to hand it is easy to envision—at least in principle—how, once a vocabulary had been created, feedback among the various brain centers involved would have allowed the children to structure their language and thought processes simultaneously. For them, what psychologists have taken to calling "private speech" would have been a conduit to the conversion of intuitions into articulated notions that could then be manipulated symbolically.

An additional attractive feature of language as the stimulus for abstract thought is that, unlike theory of mind, it is a communal possession. For the same reason that a poker player keeps his cards to himself, it would on the face of it seem to be actively disadvantageous for an individual to reveal to others his or her ability to read or accurately guess their minds. And while this reality would not obviate the spread of that ability within the population if it were merely one more expression of a generalized intelligence, in this light it is hard to see theory of mind as in itself a driver of change. Of course, we are in total *terra incognita* here; and it may even be sheer speculation that language originated as a means of communication. (After all, in a typical human paradox, it is perhaps the greatest barrier to communication that there is in the world today.) It is conceivable that the functionally important role of language as an interior conduit to thought was paramount from the beginning. But language as a means of communication would most easily and rapidly have spread through a population that possessed the necessary biology—and ultimately beyond that small original population, throughout a biologically predisposed species whose newfound intellect soon allowed it to take over the world.

LANGUAGE, SYMBOLS, AND BRAINS

It would be much easier to speculate about what happened in that leap from nonsymbolic to symbolic if we had a better idea of how the human brain works: how it turns a mass of structured electrochemical signals into what we experience as our consciousnesses. Recently developed real-time techniques of imaging what happens in the brain (i.e., where it's using energy) while it undertakes different cognitive tasks have taught us a great deal. But just how our controlling organ puts everything together into what we subjectively think and feel is still largely unknown. And this renders problematic the task of identifying those specific brain areas whose modification at the origin of our species laid the groundwork for our new cognitive performance. Nonetheless, any differences we might be able to detect between human brains and those of our close relatives would certainly give us somewhere to begin. And since paleoneurologists are still hazy about what differences in external brain shape between fossil hominids and humans mean, the natural place to start this endeavor is with the brains of living apes. After all, we have a broad idea about what we can do that the apes can't—though it turns out that there are severe practical difficulties in getting an ape to do what you want inside a functional MRI machine. Which only serves to emphasize—yet again—the vastness of the cognitive gulf that separates us from them.

We are, then, confined for the moment to static aspects of the brain in our search for the biological underpinnings of what makes us unique. And although it's long been known that there are few if any gross aspects of the human brain without their counterparts in apes, significant differences in tissue architecture are beginning to emerge as neurobiologists look at ape and human brain materials in finer and finer resolution. One recent discovery is that, while apes and humans as a group are unique in having "spindle" neurons in parts of the brain that in humans are involved in complex emotions such as trust, empathy, and guilt, we have many more of them. Scientists aren't yet sure exactly why this is, but one possible function for spindle neurons is in aiding high-speed conduction of impulses from these regions to an area at the front of the brain concerned with advance planning. It may be that the abundance of spindle neurons helps humans make swift responses to complex and changing social situations. As more and more findings such as this come to light, we will certainly be able to put together a more complete picture of what happens in human brains that doesn't in others. But for the moment, while we can be confident that the behavioral advantage humans enjoy does not simply emanate from a greater mass of brain tissue, the best we can do is make educated guesses.

My personal favorite conjecture is still the one made by the great Columbia neurobiologist Norman Geschwind back in the 1960s. Geschwind's notion was that the discrete identification of objects—naming them—was the foundation of language. Using the connection I'm making here, it would also be the foundation of symbolic cognition. In Geschwind's view, language was made possible by a physical ability to make direct associations between different areas of the brain's cortex, without passing through the older emotional centers below. The cortex is the thin sheet of neural cells that covers the outside of the brain, and it has so greatly expanded during mammal evolution (and especially in us) that, as we've seen, it has become creased and folded to fit inside the confines of the cranial vault. The largest of the folds have been used to demarcate various major functional areas of the cortex, notably its frontal, parietal (upper side), temporal (lower side), and occipital (back) "lobes." Within each lobe, other wrinkles demarcate major functional areas. Thus Broca's area, that brain region which plays a key role in motor functions including control of the speech apparatus, lies within the left frontal lobe. Modern imaging techniques have shown that many motor and other functions are in fact widely distributed through the brain; but nonetheless the major control areas identified by the great nineteenth-century neurologists are still recognized today. And most modern neuroscientists are comfortable with the idea that the prefrontal cortex, right at the forward tip of the brain, is particularly crucial in the integration of information coming in from all over this hugely complex organ. It is clearly the seat of the higher "executive" functions, coordinating and regulating the activities of the phylogenetically older parts of the brain.

But Geschwind's particular candidate for the structure permitting the associations crucial in object-naming was the "angular gyrus," a part of the parietal lobe that lies adjacent to both the temporal and the occipital areas, and is ideally positioned to mediate among all of these lobes. In humans the angular gyrus is large, whereas in all other primates it is small or missing altogether. What's more, recent imaging studies have demonstrated that it is active in the comprehension of metaphors, which are emblematic of the kind of abstract connections that are basic to language. Whether Geschwind was right or wrong, it is thus frustrating that it is next to impossible to delineate the angular gyrus in endocasts of fossil hominid brains, so we simply don't know at what point in our history it started to expand.

In figuring out just what it is that makes our brains special, we always have to keep in mind that our controlling organ is a rather untidy structure that, from very simple beginnings, has accreted rather opportunistically over an enormous period of time. So perhaps we shouldn't be looking for one single major "keystone" acquisition. Instead, the extraordinary properties of the human brain are likely emergent, resulting from a relatively tiny—and altogether accidental—addition or modification to a complex structure that was already, and exaptively, almost prepared for symbolic thought. A small tweak to an existing—and independently viable—structure gave rise to a new form of interaction among the brain's constituents that gave it an entirely unprecedented level of complexity in function.

If we can't pinpoint any specific brain *component* as the basis for our modern human consciousness, we might ask about which cognitive *systems* might have been involved. One favorite system that has been getting a lot of play recently has been our working memory, the term used by psychologists to denote our ability to hold information in the conscious mind while undertaking practical tasks. Without substantial working memory it would be impossible for us to carry out any kind of operation that involved associating several different bits of information. Proponents of the idea that working memory most crucially underwrites our complex activities do not deny that ancient hominids also needed a certain amount of such memory; but they suggest that the difference between us and them, though substantial, is one of degree, related to increasing refinement of those executive functions of the prefrontal cortex that govern decision-making, goal forming, planning, and so on. As we've seen, the various technologies that hominids developed since the inventors of stone tool making first bashed one stone with another became more complex in a highly sporadic manner. And this has been taken as evidence that working memory increased in stepwise fashion, the last major step being taken somewhere between 90 and 50 thousand years ago.

This scenario certainly fits with what the archaeological record seems to be telling us. But it still leaves the question open as to whether working memory is merely a necessary condition of our modern consciousness, rather than an altogether sufficient one. When we are pondering how we acquired our odd way of doing mental business, we might do well to consider that identifying working memory as the key ingredient to our uniqueness may in fact be analogous to selecting thermoregulation, or distance vision, or object-carrying, as the key factor that caused the earliest hominid bipeds to stand upright. The reality is that once you have the capacity concerned you have bought an entire package of advantages, plus any disadvantages that might come along with it. In the case of bipedality, standing upright was almost certainly just what came naturally to the creatures involved. In the case of symbolic consciousness, it seems likely that a random modification of the already exapted brain, plus some children at play, led to the literal emergence of a phenomenon that changed the world.

CODA

None of what I said at the end of chapter 14 implies that our species has necessarily changed our planet in an intentional way. It's certainly reasonable to suppose that there was no such intention at the start, in a world where our forebears were hunters and gatherers: people who were largely, if not entirely, integrated into their ecosystems. It is, though, highly probable that from the very beginning, apart from death, the only ironclad rule of human experience has been the Law of Unintended Consequences. Our brains are extraordinary mechanisms, and they have allowed us to accomplish truly amazing things; but we are still only good at anticipating—or at least of paying attention to—highly immediate consequences. We are notably bad at assessing risk, especially long-term risk. We believe crazy things, such as that human sacrifice will propitiate the gods, or that people are kidnapped by space aliens, or that endless economic expansion is possible in a finite world, or that if we just ignore climate change we won't have to face its consequences. Or at the very least, we act as if we do.

All of this is, of course, in perfect agreement with the untidy accretionary history of the human brain. Inside our skulls are fish, reptile, and shrew brains, as well as the highest centers that allow us to integrate information in our unique way; and some of our newer brain components talk to each other via some very ancient structures indeed. Our brains are makeshift structures, opportunistically assembled by Nature over hundreds of millions of years, and in multiple different ecological contexts. When we realize that our symbolic capacities are an incredibly recent acquisition—and not merely the icing on the cake, but the candy bead surmounting the cherry atop the icing—it becomes evident that our brains as they perform now cannot have been fine-tuned by evolution for anything. We have achieved our mental eminence only because a long series of ancestors, stretching back into the most remote reaches of time, happened simply to have been able to cope with prevailing circumstances better than their competitors could. And then a final, still inscrutable, acquisition just happened to make a huge difference. If anything had occurred otherwise, anywhere along that lengthy trail, you wouldn't be reading this book today.

There is a school of thought that we humans sometimes act so bizarrely because the evolution of our brains has not been able to keep pace with the rapid transformation of society that has occurred since humans began to adopt settled lifestyles at the end of the last glacial episode. Our minds, according to this view, are still responding, sometimes inappropriately, to the exigencies of a bygone "environment of evolutionary adaptedness." This view has a wonderful reductionist appeal; but in reality our brains are the ultimate general-purpose organs, not adapted "for" anything at all. Yes, you can indeed find regularities in human behaviors, every one of them doubtless limited by basic commonalities in the structure of our controlling organs. But all such regularities are in reality statistical abstractions, and people are absolutely uniform in none of them. As a result, if any statistical phenomenon could be said to govern the human condition, it would be the "normal distribution," or the "bell curve." This describes the frequency with which different expressions of the same feature occur within a population. Tall in the center, where most individuals cluster, this bell-shaped curve tails off more or less symmetrically to each side, reflecting the fact that most observations of any feature fall close to the average, with deviations becoming increasingly rare the farther they lie from the mean.

In any human characteristic you might care to specify, physical or behavioral, you will find a bell curve. Only a few of us are very smart, or very dumb; most of us are somewhere in the middle. Ditto with tall/short, caring/indifferent, strong/weak, chaste/promiscuous, spiritual/profane, or any other continuous variable you might care to mention. This is, of course, why the internal human condition is virtually impossible to pin down: you can easily find an individual *Homo sapiens* to exemplify each extreme of any behavioral spectrum you can dream up.

For every saint, there is a sinner; for every philanthropist, a thief; for every genius, an idiot. In this perspective, having bad people around is simply the price we pay for having good ones as well. Put another way, there is no need to look for special explanations for altruism when this feature is matched on the other side of the curve by selfishness. Individuals themselves are typically bundles of paradoxes, each of us mixing admirable with less worthy traits, even expressing the same trait differently at different times. We are ruled by our reason, but only until our hormones take over.

Similarly, nobody subscribes to *all* of those crazy ideas that are floating around, though most of us are attracted to a few of them. One of those crazy ideas is that the human condition can somehow be described by a long laundry list of "human universals"—uniquely human psychological and behavioral features that everybody has. But it nearly always turns out that these "universals" are either not uniquely human, or not universal among humans. Indeed, apart from that basic ability we all share to re-create the world in the mind, perhaps the only other true "human universal" we all show is cognitive dissonance.

Because of its peculiar cognitive properties, our species and its individual members are entities of entirely different kinds. For, while individual human beings are substantially—though not entirely—the products of their own particular genomes, coming into the world as broadly the kind of persons they will be as adults, this is not true in the same way for the human species as a whole. Indeed, the universal human condition will always remain elusive (and incessantly debated) for the very good reason that it is inherently unspecifiable.

So, what are we to make of ourselves? After a long evolutionary history we have arrived at a point at which our accidental cognitive prowess is allowing us unwittingly to change the very surface of the Earth on which we live. Indeed, it has recently been proposed (with typical human arrogance) that we should rename the current Holocene epoch of geological time the "Anthropocene" (roughly, the "new human age"). Many geologists cringe at this suggestion (the invention of an ecologist and an atmospheric chemist) because the depredations of a single species have never been used as a criterion for defining a phase of geological time. Nonetheless, it is nothing short of alarming how human intervention is affecting a huge array of processes that will clearly be reflected in the record available to future geologists, should there be any. To take just one example of many, over the eons all the natural might of the elements has typically lowered the surfaces of the continents by a few tens of meters per million years. In shocking contrast, a recent analysis has shown that trends in human activity that started only around the beginning of the first millennium have led to a current rate of worldwide erosion that is ten times higher.

If this were simply another statistical abstraction, maybe it wouldn't matter much to members of such a supremely inward-looking species as our own. But we humans are already reaping enormous practical consequences of all that removal of the continental crust, in terms both of vastly accelerated inland erosion, and of coastal sedimentation. We are literally reshaping the planet on which we live. And behaviors that a resilient environment could simply absorb when *Homo sapiens* was thin on the ground become hugely damaging to human populations when there are seven billion of us around. What's more, the effects cut both ways: the more enormous and complex the human enterprise grows, the more fragile it becomes. A flood that would have been a mere inconvenience to scattered groups of hunter-gatherers becomes a human tragedy of major dimensions in the crowded landscapes of Bangladesh or the Mississippi Valley. In many unintentional as well as intended ways we have indeed shown ourselves to be Masters of the Planet; but that is no guarantee that the Planet will not bite back when overstressed.

Clearly, then, we are already threatened by some of the correlates of those very attributes that make us so remarkable. And this makes it natural, of course, to ask whether we are condemned by the evolutionary "design" of our brains to continue in our self-destructive ways. Fortunately, the answer is "no"—at least in principle. For example, while it has been shown that certain tendencies to violence originating in unusual brain activity may well be inherited, it is also known that those tendencies can usually be modified by environment and experience. What is more, societies as wholes are not necessarily subject to the same imperfections as individuals are—though, sadly, this cannot be said of their leaders. Indeed, complex societies, with all their rules and regulations, bizarre procedures, and sometimes draconian means of coercement, exist most usefully to compensate for the manifold deficiencies of individuals, especially those whose behaviors lie toward the negative extremes of some of those many bell curves. It is not least because of the often grating social and legal strictures that govern our existences that most of us behave reasonably responsibly most of the time; and it may not be hopelessly unrealistic to suggest that, once the gravity of our situation becomes clear to a majority, societies may actually be capable of taking the hard decisions that will be necessary if we are to maintain an equilibrium with the planet that supports us.

Still, while we are not perfected beings, we have nonetheless come a long way in the past seven million years. Does this mean that we can passively wait for evolution to complete its job? With a bit of patience, will the workings of natural selection eventually make us cleverer, and better aware of the full range of consequences of our actions? Unfortunately, this time the answer is also no, or at least not if present demographic trends continue. Our ancestors evolved during a time when hominids were thinly scattered over the landscape, in tiny and therefore genetically unstable populations that were subjected to frequent environmental stresses and disruptions. These were ideal conditions in which new populations and species could emerge and incorporate significant genetic and physical innovations. Indeed, more than likely it was just those highly unsettled prevailing circumstances that, in combination with our cultural proclivities, accounted for the unusually fast tempo of hominid evolution during the Pleistocene.

But that was then, and this is now. Since our adoption of settled existences at the end of the last Ice Age, the human population has mushroomed, until we are now packed across the globe with precious little elbow room left. These new conditions have changed the rules of the game entirely. Modern human populations have simply become too large and dense to witness the fixation of any significant genetic novelties that might in theory make us smarter and more protective of our own long-term interests. Short of a dramatic change in our demographic circumstances, we are stuck with our murky selves.

Many will find this prospect substantially less than optimal; but luckily it is far from the entire story. This is because, while the prospects for biological improvement seem rather dim in the absence of some (easily imaginable) cataclysm that might re-establish normal evolutionary rules, human innovation in the broader sense has not run into a brick wall. There is no question that our cognitive as well as our anatomical systems are far from perfect in a whole variety of areas. But our rational abilities and our extravagant neophilia nonetheless remain beyond remarkable. From the very first stirrings of the human symbolic spirit, the technological and creative histories of humankind have revolved around an energetic exploration of the innovative potential released by our new way of processing information about the world. And if one thing is clear above all, it is that this exploration of our existing capacity is far from exhausted. Indeed, one might even argue that it has barely begun. So, while the auguries appear indeed to be for no significant biological change in our species, culturally, the future is infinite.

ACKNOWLEDGMENTS

This book is the distillation of a longish career during which I have learned from, and been influenced by, many colleagues. They are too many to cite by name, but all know who they are, and all have my deepest appreciation. Still, while for me personally it is something of a culmination, in a larger sense this volume is merely a progress report. Science is a moving target, and nothing will give me deeper pleasure than to see the ideas I have expressed in these pages overtaken by developments in a fast-moving field. What we were taught when I came into paleoanthropology almost fifty years ago looks engagingly quaint today; and I have no doubt that today's state of the art will look equally odd a half-century from now.

My appreciation goes to Amir Aczel for introducing me to my editor at Palgrave Macmillan, Luba Ostashevsky. Luba encouraged me to start this book, and steadfastly saw it through. Also at Palgrave Macmillan I'd like to thank Laura Lancaster and Donna Cherry for holding my hand through the production process, and Ryan Masteller for his copyedit. And it was fun working with Christine Catarino and Siobhan Paganelli.

I will always be grateful to Jane Isay and Michelle Press for showing me how rewarding it can be to write for a general audience. Thanks, too, to all those photographers and artists, acknowledged in the captions, who produced the illustrations, with a special nod to Jay Matternes and Jenn Steffey. And my gratitude goes above all to my wife Jeanne, for her support and forbearance during all that writing.

NOTES AND
BIBLIOGRAPHY

Immediately below is a short list of recent books that may be of interest to those with a desire to read further about human evolution. The coverage of each work will be evident from its title; especially well illustrated are Johanson and Edgar (2006), Sawyer et al. (2007), and Tattersall and Schwartz (2000). For a comprehensive bibliography covering all the topics mentioned in this book, see Tattersall (2009), which also contains the history of discovery and ideas in paleoanthropology that is largely neglected here. For the widest possible coverage at a semi-technical level of topics relating to hominid paleontology and Paleolithic archaeology, Delson et al. (2000) is highly recommended. Following this general list is a chapter-by-chapter bibliography identifying the main primary sources consulted for this book, and all works from which quotations are made.

Delson, E., I. Tattersall, J. A. Van Couvering, A. S. Brooks. 2000. *Encyclopedia of Human Evolution and Prehistory*, 2nd. ed. New York: Garland Press.

DeSalle, R., I. Tattersall. 2008. *Human Origins: What Bones and Genomes Tell Us About Ourselves*. College Station, TX: Texas A&M University Press.

Eldredge, N. 1995. *Dominion*. New York: Henry Holt.

Gibbons, A. 2006. *The First Human: The Race to Discover Our Earliest Ancestors*. New York: Doubleday.

Hart, D., R. W. Sussman. 2009. *Man the Hunted: Primates, Predators, and Human Evolution*. Expanded ed. New York: Westview/Perseus.

Johanson, D. C., B. Edgar. 2006. *From Lucy to Language*, 2nd ed. New York: Simon and Schuster.

Klein, R. 2009. *The Human Career: Human Biological and Cultural Origins*, 3rd ed. Chicago: University of Chicago Press.

Klein, R., B. Edgar. 2002. *The Dawn of Human Culture*. New York: Wiley.

Sawyer, J. G., V. Deak, and E. Sarmiento. 2007. *The Last Human: A Guide to Twenty-Two Species of Extinct Humans*. New Haven, CT: Yale University Press.

Stringer, C. B., P. Andrews. 2005. *The Complete World of Human Evolution*. London and New York: Thames and Hudson.

Tattersall, I. 2009. *The Fossil Trail: How We Know What We Think We Know about Human Evolution*, 2nd ed. New York: Oxford University Press.

Tattersall, I. 2010. *Paleontology: A Brief History of Life.* Consohocken, PA: Templeton Foundation Press.

Tattersall, I., J. H. Schwartz. 2000. *Extinct Humans.* New York: Westview/Perseus.

Wade, N. 2006. *Before the Dawn: Recovering the Lost History of Our Ancestors.* New York: Penguin Press.

Wells, S. 2007. *Deep Ancestry: Inside the Genographic Project.* Washington, DC: National Geographic.

Zimmer, C. 2005. *Smithsonian Intimate Guide to Human Origins.* New York: HarperCollins.

CHAPTER 1: ANCIENT ORIGINS

For an accessible recent account of Rift Valley formation and the early East African hominoids, see Walker and Shipman (2005). Pickford (1990) discusses eastern African uplift in relation to hominoid evolution, and Harrison (2010) provides an excellent overview of the variety and relationships of fossil hominoids and putative hominid precursors. For more about *Oreopithecus* see Köhler and Moyà-Solà (1997), Moyà-Solà et al. (1999), and Rook et al. (1999). *Pierolapithecus* was described by Moyà-Solà et al. (2004). For a history of the recognition of hominids and key criteria see Tattersall (2009). *Sahelanthropus* was described by Brunet et al. (2002, 2005), and virtually reconstructed by Zollikofer et al. (2005). *Orrorin* and its environment were described by Senut et al. (2001) and Pickford et al. (2001, 2002). *Ardipithecus ramidus* was named (initially as *Australopithecus ramidus*) by White et al. (1994), and its skeleton comprehensively analyzed in a special issue of *Science* (White et al., 2009). *Ardipithecus kadabba* was described by Haile-Selassie (2001) and Haile-Selassie et al. (2004). For more on bipedality see Harcourt-Smith (2007). *Australopithecus anamensis* was first described by Leakey et al. (1995, 1998), and the Kenya material was comprehensively presented by Ward et al. (2001). Claimed Ethiopian material of this species was presented by White et al. (2006), and gradual transformation of *A. anamensis* into *A. afarensis* was advocated by Kimbel et al. (2006).

Brunet, M., F. Guy, D. Pilbeam, H. T. Mackaye, A. Likius, D. Ahounta, A. Beauvilain, C. Blondel, H. Bocherens, J.-R. Boisserie, L. De Bonis, Y. Coppens, J. Dejax, C. Denys, P. Duringer, V. Eisenmann, G. Fanone, P. Fronty, D. Geraads, T. Lehmann, F. Lihoreau, A. Louchart, A. Mahamat, G. Merceron, G. Mouchelin, O. Otero, P. P. Campomanes, M. Ponce de León, J.-C. Rage, M. Sapanet, M. Schuster, J. Sudre, P. Tassy, X. Valentin, P. Vignaud, L. Viriot, A. Zazzo, C. Zollikofer. 2002. A new hominid from the Upper Miocene of Chad, Central Africa. Nature: 145–151.

Brunet, M., F. Guy, D. Pilbeam, D. E. Lieberman, A. Likius, H. T. Mackaye, M. S. Ponce de León, C. P. E. Zollikofer, P. Vignaud. 2005. New material of the earliest hominid from the Upper Miocene of Chad. *Nature* 434: 752–754.

Haile-Selassie, Y. 2001. Late Miocene hominids from the Middle Awash, Ethiopia. *Nature* 412: 178–181.

Haile-Selassie, Y., G. Suwa, and T. D. White. 2004. Late Miocene teeth from Middle Awash, Ethiopia, and early hominid dental evolution. *Science* 303: 1503–1505.

Harcourt-Smith, W. E. H. 2007. The origins of bipedal locomotion. In *Handbook of Paleoanthropology, Volume 3*. W. Henke and I. Tattersall, eds. Heidelberg and New York: Springer, 1483–1518.

Harrison, T. 2010. Apes among the tangled branches of human origins. *Science* 327: 532–534.

Keith, A. 1915. *The Antiquity of Man*. London: Williams and Norgate.

Kimbel, W. H., C. A. Lockwood, C. V. Ward, M. G. Leakey, Y. Rak, D. Johanson. 2006. Was *Australopithecus anamensis* ancestral to *A. afarensis?* A case of anagenesis in the hominin fossil record. *Jour Hum. Evol.* 51: 134–152.

Köhler, M., S. Moyà-Solà. 1997. Ape-like or hominid-like? The positional behavior of *Oreopithecus* reconsidered. *Proc. Nat. Acad. Sci. USA* 94: 11747–11750.

Leakey, M. G., C. S. Feibel, I. McDougall, C. Ward, A. Walker. 1995. New four-million-year-old hominid species from Kanapoi and Allia Bay, Kenya. *Nature* 376: 565–571.

Leakey, M. G., C. S. Feibel, I. McDougall, C. Ward, A. Walker. 1998. New specimens and confirmation of an early age for *Australopithecus anamensis*. *Nature* 393: 62–66.

Moyà-Solà, S., M. Köhler, L. Rook. 1999. Evidence of hominid-like precision grip capability in the hand of the Miocene ape *Oreopithecus*. *Proc. Nat. Acad. Sci. USA* 96: 313–317.

Moyà-Solà, S., M. Köhler, D. M. Alba, I. Casanova-Vilar, J. Galindo. 2004. *Pierolapithecus catalaunicus*, a new Middle Miocene great ape from Spain. *Science* 306: 1339–1344.

Pickford, M. 1990. Uplift of the roof of Africa and its bearing on the origin of mankind. *Hum. Evol.* 5: 1–20.

Pickford, M. and Senut B. 2001. The geological and faunal context of Late Miocene hominid remains from Lukeino, Kenya. *C. R. Acad. Sci. Paris,* ser. IIa, 332: 145–152.

Pickford, M., B. Senut, D. Gommery, J. Treil. 2002. Bipedalism in *Orrorin tugensis* revealed by its femora. *C. R. Palévol.* 1: 191–203.

Rook, L., L. Bondioli, M Köhler, S. Moyà-Solà, R. Macchiarelli. 1999. *Oreopithecus* was a bipedal ape after all: Evidence from the iliac cancellous architecture. *Proc. Nat. Acad. Sci. USA* 96: 8795–8799.

Senut, B., M. Pickford, D. Gommery, P. Mein, K. Cheboi, Y. Coppens. 2001. First hominid from the Miocene (Lukeino Formation, Kenya). *C. R. Acad. Sci. Paris,* ser. IIa, 332: 137–144.

Tattersall, I. 2009. *The Fossil Trail: How We Know What We Think We Know about Human Evolution*. 2nd ed. New York: Oxford University Press.

Walker, A., P. Shipman. 2005 *The Ape in the Tree: An Intellectual and Natural History of* Proconsul. Harvard: Belknap Press.

Ward, C. V., M. G. Leakey, A. Walker. 2001. Morphology of *Australopithecus anamensis* from Kanapoi and Allia Bay, Kenya. *Jour. Hum. Evol.* 41: 255–368.

White, T. D., G. WoldeGabriel, B. Asfaw, S. Ambrose, Y. Bayene, R. L. Bernor, J.-R. Boisserie, and numerous others. 2006. Assa Issie, Aramis and the origin of *Australopithecus*. *Nature* 440: 883–889.

White, T. D. and numerous others. 2009. Special Issue on *Ardipithecus ramidus*. *Science* 326: 5–106.

Zollikofer, C. P. E., M. S. Ponce de León, D. E. Lieberman, F. Guy, D. Pilbeam, A. Likius, H. T. Mackaye, P. Vignaud, M. Brunet. 2005. Virtual cranial reconstruction of *Sahelanthropus tchadensis*. *Nature* 434: 755–759.

CHAPTER 2: THE RISE OF THE BIPEDAL APES

The classic description of the early Hadar hominid collections is found in Johanson et al. (1982), and the Hadar *Australopithecus afarensis* crania and other more recently collected specimens are documented by Kimbel et al. (2004). A captivating general account of the discovery and initial analysis of the *A. afarensis* fossils is by Johanson and Edey (1982). It is still in print. See Aronson et al. (2008) for an overview of Hadar environments. The "hyper-bipedal" interpretation of *A. afarensis* locomotion is summarized by Lovejoy (1988); the classic reinterpretation emphasizing arboreal features is by Stern and Susman (1983), and limb proportions were analyzed by Jungers (1982). Rak (1991) reconsidered Lucy's pelvic anatomy, and a recent overview of *A. afarensis* locomotion is provided by Ward (2002). For dental descriptions of modern hominoids see Aiello and Dean (1990), and of *A. afarensis* see Johanson and White (1979). A recent analysis of dental microwear in the latter is provided by Ungar (2004). For an account of the A. L. 333 locality, see Behrensmeyer (2008). An overall description of the Laetoli sites is found in Leakey and Harris (1987), and the most recent analysis of the footprints is by Raichlen et al. (2010). *Australopithecus afarensis* was named by Johanson et al. (1978). The original account of the Dikika child was by Alemseged et al. (2006), and it is well illustrated in Sloan (2006). The cut-marked bones from Dikika were published by McPherron et al. (2010), and the Woranso-Mille skeleton by Haile-Selassie et al. (2010). The Bouri hominid was named by Asfaw et al. (1999), and the cut-marked bones from the same deposits were published by deHeinzelin et al. (1999). The Gona stone tools were announced by Semaw (2000), and the cut-marked bones from this area by Dominguez-Rodrigo (2005). The Kanzi research was reported by Schick et al. (1999).

Aiello, L., C. Dean. 1990. *An Introduction to Human Evolutionary Anatomy.* London and San Diego: Academic Press.

Alemseged, Z., F. Spoor, W. H. Kimbel, R. Bone, D. Geraads, D. Reed, J. G. Wynn. A juvenile early hominid skeleton from Dikika, Ethiopia. *Nature* 443: 296–301.

Aronson, J. L., M. Hailemichael, S. M. Savin. 2008. Hominid environments at Hadar from paleosol studies in a framework of Ethiopian climate change. *Jour. Hum. Evol.* 55: 532–550.

Asfaw, B., T. White, O. Lovejoy, B. Latimer, S. Simpson and G. Suwa. 1999. *Australopithecus garhi*: A new species of early hominin from Ethiopia. *Science* 284: 629–635.

Behrensmeyer, A. K. 2008. Paleoenvironmental context of the Pliocene A.L. 333 "First Family" hominin locality, Hadar Formation, Ethiopia. *Geol. Soc. Amer. Spec. Pap.* 446: 203–235.

deHeinzelin, J., J. D. Clark, T. White, W. Hart, P. Renne, G. WoldeGabriel, Y. Beyene, E. Vrba. 1999. Environment and Behavior of 2.5-million-year-old Bouri hominids. *Science* 284: 625–629.

Dominguez-Rodrigo, M., T. R. Pickering, S. Semaw, M. J. Rogers. 2005. Cut-marked bones from Pliocene archaeological sites at Gona, Ethiopia: Implications for the function of the world's earliest stone tools. *Jour. Hum. Evol.* 48: 109–121.

Haile-Selassie, Y, B. M. Latimer, M. Alene, A. L. Deino, L. Gibert, S. M. Melillo, B. Z. Saylor, G. R. Scott, and C. O. Lovejoy. 2010. An early *Australopithecus afarensis* postcranium from Woranso-Mille, Ethiopia. *Proc. Nat. Acad. Sci. USA* 107: 12121–12126.

Johanson, D. C., M. Edey. 1982: *Lucy: The Beginnings of Humankind*. New York: Warner Books.

Johanson, D. C., T. White. 1979. A systematic assessment of early African hominids. *Science* 203: 321–330.

Johanson, D. C., T. D. White, Y. Coppens. 1978. A new species of the genus *Australopithecus* (Primates: Hominidae) from the Pliocene of eastern Africa. *Kirtlandia* 28: 1–14.

Johanson, D. C., et al. 1982. Special Issue: Pliocene hominid fossils from Hadar, Ethiopia. *Amer. Jour. Phys. Anthropol.* 57: 373–724.

Jungers, W. L. Lucy's limbs: Skeletal allometry and locomotion in *Australopithecus afarensis*. *Nature* 297: 676–678.

Kimbel, W. H., Y. Rak, D. C. Johanson. 2004. *The Skull of* Australopithecus afarensis. Oxford and New York: Oxford University Press.

Leakey, M. D., J. M. Harris (eds.). 1987. *Laetoli: A Pliocene Site in Northern Tanzania*. Oxford: Clarendon Press.

Lovejoy, C. O. 1988. Evolution of human walking. *Scientific American* 259: 118–125.

McPherron, S., Z. Alemseged, C. W. Marean, J. G. Wynne, D. Reed, D. Geraads, R. Bobe, H. A. Béarat. 2010. Evidence for stone-tool-assisted consumption of animal tissues before 3.39 million years ago at Dikika, Ethiopia. *Nature* 466: 857–860.

Raichlen, D. A., A. D. Gordon, W. E. H. Harcourt-Smith, A. D. Foster, W. R. Haas. 2010. Laetoli footprints preserve earliest direct evidence of human-like bipedal biomechanics. *PLoS One* 5 (3): e9769.

Rak, Y. 1991. Lucy's pelvic anatomy: its role in bipedal gait. *Jour. Hum. Evol.* 20: 283–290.

Schick, K., N. Toth, G. Garufi, E. S. Savage-Rumbaugh, D. Rumbaugh, R. Sevcik. 1999. Continuing investigations into the stone tool-making and tool-using capabilities of a bonobo (*Pan paniscus*). *Jour. Archaeol. Sci.* 26: 821–832.

Semaw, S. 2000. The world's earliest stone artifacts from Gona, Ethiopia: Their implications for understanding stone technology and patterns of human evolution between 2.6–1.5 million years ago. *Jour. Archaeol. Sci.* 27: 1197–1214.

Stern, J. T., R. L. Susman. 1983. The locomotor anatomy of *Australopithecus afarensis*. *Amer. Jour. Phys. Anthropol.* 60: 279–317.

Ungar, P. 2004. Dental topography and diets of *Australopithecus afarensis* and early *Homo*. *Jour. Hum. Evol.* 46: 605–622.

Ward, C. V. 2002. Interpreting the posture and locomotion of *Australopithecus afarensis*: Where do we stand? *Yrbk Phys. Anthropol.* 45: 185–215.

CHAPTER 3: EARLY HOMINID LIFESTYLES AND THE INTERIOR WORLD

The cooking hypothesis is most comprehensively presented by Wrangham (2009), and the tapeworm research is by Hoberg et al. (2001); possible hypervitaminosis A in a fossil hominid is reported by Walker et al. (1982); and scavenging and early hominid social organization are discussed by Hart and Sussman (2009). Leopard-kill stealing is suggested by Cavallo and Blumenschine (1989). Stable isotope research on South African australopiths is summarized by Sponheimer and Lee-Thorp (2007), and East African *Paranthropus* isotopic analyses were reported by Cerling et al. (2011). Chimpanzee scavenging frequency was reported by Watts (2008), and spear-hunting at Fongoli by Pruetz and Bertolani (2007). Stanford (1999) and Mitani and Watts (2001) provide overviews of chimpanzee hunting behaviors; and Gomes and Boesch (2009) discuss meat-sharing and sex among chimpanzees. Use and antiquity of stone anvils by chimpanzees is discussed by Mercader et al. (2007), and power scavenging in contributions to Stanford and Bunn (2001). Calvin (1996) provides an accessible account of throwing and associated neural mechanisms. Dart quotation is from Dart (1953). For an overview and bibliography of cognitive issues see Tattersall (2011); and for mirror self-recognition see Gallup (1970). Seyfarth and Cheney (2000; quote from p. 902) reported cognitive results on monkeys; Povinelli observations and quotes come from Povinelli (2004: 33, 34).

Calvin, W. H. 1996. *How Brains Think: Evolving Intelligence, Then and Now.* New York: Basic Books.

Cavallo, J. A., R. J. Blumenschine. 1989. Tree-stored leopard kills: expanding the hominid scavenging niche. *Jour. Hum. Evol.* 18: 393–400.

Cerling, T. E., E. Mbua, F. M. Kirera, F. K. Manthi, F. E. Grine, M. G. Leakey, M. Sponheimer, K. T. Uno. 2011. Diet of *Paranthropus boisei* in the early Pleistocene of East Africa. *Proc. Nat Acad. Sci. USA* 108: 9337–9341.

Dart, R. A. 1953. The predatory transition from ape to man. *Intl Anthopol. Ling. Rev.* 1: 201–217.

Gallup, G. G. 1970. Chimpanzees: Self-recognition. *Science* 167: 86–87.

Gomes, C. M., C. Boesch. 2009. Wild chimpanzees exchange meat for sex on a long-term basis. *PLoS One* 4: e5116.

Hart, D., R. W. Sussman. 2009. *Man the Hunted: Primates, Predators, and Human Evolution.* Expanded edition. Boulder, CO: Westview Press.

Hoberg, E. P., N. L. Alkire, A. de Queiroz, A. Jones. 2001. Out of Africa: Origins of the *Taenia* tapeworms. *Proc. Roy. Soc. Lond. B.* 268: 781–787.

Mercader, J., H. Barton, J. Gillespie, J. Harris, S. Kuhn, R. Tyler, and C. Boesch. 2007. 4,300-year-old chimpanzee sites and the origins of percussive stone technology. *Proc. Nat. Acad. Sci. USA* 104: 3043–3048.

Mitani, J. C., D. P. Watts. Why do chimpanzees hunt and share meat? *Anim. Behav.* 61: 915–924.

Povinelli, D. J. 2004. Behind the ape's appearance: Escaping anthropocentrism in the study of other minds. *Daedalus* 133 (1): 29–41.

Pruetz, J. D., P. Bertolani. Savanna chimpanzees, *Pan troglodytes verus*, hunt with tools. *Curr. Biol.* 17: 412–417.

Seyfarth, R. M., Cheney, D. L. 2000. Social awareness in monkeys. *Amer. Zool.* 40: 902–909.

Sponheimer, M., J. Lee-Thorp. 2007. Hominin paleodiets: The contribution of stable isotopes. In W. Henke and I. Tattersall (eds,), *Handbook of Paleoanthropology.* Heidelberg: Springer, 555–585.

Stanford, C. B. 1999. *The Hunting Apes: Meat-eating and the Origins of Human Behavior.* Princeton: Princeton University Press.

Stanford, C. B. H. Bunn. 2001. *Meat-eating and Human Evolution.* New York: Oxford University Press.

Tattersall, I. 2011. Origin of the human sense of self. In W. van Huyssteen and
  E. B. Wiebe (eds.), *In Search of Self*. Chicago: Wm. B. Eerdmans, 33–49.
Walker, A. C., M. R. Zimmerman, R. E. F. Leakey. 1982. A possible case of hy-
  pervitaminosis A in *Homo erectus*. *Nature* 296: 248–250.
Watts, D. 2008. Scavenging by chimpanzees at Ngogo and the relevance of chim-
  panzee scavenging to early hominid behavioral ecology. *Jour. Hum. Evol.*
  54: 125–133.
Wrangham, R. 2009. *Catching Fire: How Cooking Made Us Human*. New
  York: Basic Books.

CHAPTER 4: AUSTRALOPITH VARIETY

For the latest dating of the South African australopith sites see Herries et al.
(2009); South African australopith morphologies are reviewed in various contributions in Grine (1988). The most recent account of the Little Foot skeleton
is by Clarke (2008). *Australopithecus sediba* was described by Berger et al.
(2010). For dental microwear studies see Scott et al. (2005) and Ungar et al.
(2008), and for a review and analysis of stable carbon isotope results see Sponheimer and Lee-Thorp (2007). For an overview of South African Early Stone
Age tools, see Kuman (2003), and for the manipulatory abilities of Swartkrans
hominids, see Susman (1994). The classic account of the robust "Zinjanthropus" from Olduvai Gorge is by Tobias (1967); *Homo habilis* was named by
L. Leakey, Tobias, and Napier (1964); and the Ethiopian Omo Basin hominids
were summarized by Howell (1978). Wood (1991) gives an account of the East
Turkana hominids. The Black Skull was described by Walker et al. (1986), and
the Konso skull by Suwa et al. (1997). Wood and Collard (1999) reconsidered
the allocation to genera of early hominids, and M. G. Leakey et al. (2001)
described *Kenyanthropus*.

Berger, L. R., D. J. de Ruiter, S. E. Churchill, P. Schmid, K. J. Carlson, P. H. G.
  M. Dirks, J. M. Kibii. 2010. *Nature* 328: 195–204.
Clarke, R. J. 2008. Latest information on Sterkfontein's *Australopithecus* skel-
  eton and a new look at *Australopithecus*. *S. Afr. Jour. Sci.* 104: 443–449.
Grine, F. E. (ed). 1988. *Evolutionary History of the "Robust" Australopith-
  ecines*. Hawthorne, NY: Aldine de Gruyter.
Herries, A. I. R., D. Curnoe, J. W. Adams. 2009. A multi-disciplinary seriation
  of early *Homo* and *Paranthropus* bearing palaeocaves in southern Africa.
  Quat. Int. 202: 14–28.
Howell, F. C. 1978. Hominidae. In V. J. Maglio and H. B. S. Cooke (eds.). *Evolution of African Mammals*. Cambridge, MA: Harvard University Press,
  154–248.
Kuman, K. 2003. Site formation in the early South African Stone Age sites and
  its influence on the archaeological record. *S. Afr. Jour. Sci.* 99: 251–254.
Leakey, L. S. B., P. V. Tobias, J. R. Napier. 1964. A new species of genus *Homo*
  from Olduvai Gorge. *Nature* 202: 7–9.
Leakey, M. G., F. Spoor, F. H. Brown, P. N. Gathogo, L. N. Leakey, I. McDougall. 2001. New hominin genus from eastern Africa shows diverse middle
  Pliocene lineages. *Nature* 410: 433–440.
Scott, R. S., P. S. Ungar, T. S. Bergstrom, C. A. Brown, F. E. Grine, M. F. Teaford,
  A. Walker. 2005. Dental microwear texture analysis shows within-species
  diet variability in fossil hominins. *Nature* 436: 693–695.

Sloan, C. P. 2006. The origin of childhood. *National Geographic* 210 (5): 148–159.

Susman, R. L. 1994. Fossil evidence for early hominid tool use. *Science* 265: 1570–1573.

Suwa, G., B. Asfaw, Y. Beyene, T. D. White, S. Katoh, S. Nagaoka, H. Nakaya, K. Uzawa, P. Renne, G. WoldeGabriel. 1997. The first skull of *Australopithecus boisei. Nature* 389: 489–446.

Tobias, P. V. 1967. *Olduvai Gorge,* Vol. 2. Cambridge: Cambridge University Press.

Ungar, P., F. E. Grine, M. F. Teaford. 2008. Dental microwear and diet of the Plio-Pleistocene hominin *Paranthropus boisei. PLoS One* 3: e2044.

Walker, A. C., R. E. F. Leakey, J. M. Harris, F. H. Brown. 1986. 2.5-Myr *Australopithecus boisei* from west of Lake Turkana, Kenya. *Nature* 322: 517–522.

Wood, B. 1991. *Koobi Fora Research Project,* Vol. 4. Oxford: Clarendon Press.

Wood, B., M. Collard. The human genus. *Science* 284: 65–71.

CHAPTER 5: STRIDING OUT

The "Man the Toolmaker" concept was most widely popularized by Kenneth Oakley (1949 and many subsequent editions). Louis Leakey et al. (1964) named *Homo habilis,* and the similarities of its type material to australopiths was noted by, among others, Robinson (1965) and Pilbeam and Simons (1965). KNM-ER 1470 was first described (simply as a member of *Homo*) in 1973 by R. E. F. Leakey, who by 1976 was calling it *Homo habilis.* It was allocated to *Homo rudolfensis* by Alexeev (1986), and transferred again to *Kenyanthropus* by M. G. Leakey et al. (2001). See Tattersall (2009) for the history of other hominids allocated to *Homo habilis, Homo rudolfensis,* and "early *Homo,*" and see Schwartz and Tattersall (2005) for morphological discussion of these fossils. Dobzhansky's views on variability among early hominids were first published in 1944, and May's influential Cold Spring Harbor paper in 1950. For a discussion of the Evolutionary Synthesis and its sequelae in evolutionary biology, see Eldredge (1985); specifically in paleoanthropology, see Tattersall (2009).

Wood and Collard (1999) reappraised the content of the genus *Homo.* Dubois described *Pithecanthropus erectus* most fully in 1894; see Schwartz and Tattersall (2005) and Tattersall (2007) for a full discussion of the *Homo erectus/ Homo ergaster* issue. KNM-WT 15000 was described and analyzed most comprehensively in the various contributions to Walker and Leakey (1993). MacLarnon and Hewitt (1999) reviewed the wider significance of the vertebral canal in breathing control. Broca's area has been recently reappraised in some detail by Amunts et al. (2010). Growth and life history features of the Turkana Boy were reappraised by Dean et al. (2001), Dean and Smith (2009), and Graves et al. (2010), and the East Turkana footprints were reported by Bennett et al. (2009). The significance of brain size in a juvenile *Homo erectus* was analyzed by Coqueugniot et al. (2004). Goldschmidt (1940) published the notion of the "hopeful monster," and Peichel et al. (2001) presented results on gene regulation in sticklebacks. Gene expression in tissues of chimpanzees and humans was reported by Khaitovich et al. (2005).

Alexeev, V. P. 1986. *The Origin of the Human Race.* Moscow: Progress Publishers.

Amunts, K., M. Lenzen, A. D. Friederici, A. Schleicher, P. Morosan, N. Palomero-Gallagher, K. Zilles. 2010. Broca's region: Novel organization principles and multiple receptor mapping. *PLoS Biol.* 8: e1000489.

Bennett, M. R., J. W. K. Harris, B. G. Richmond, D. R. Braun, E. Mbua, P. Kiura, D. Olago, M. Kibunjia, C. Omuombo, A. K. Behrensmeyer, D. Huddart, S. Gonzalez. 2009. Early hominin foot morphology based on 1.5 million-year-old footprints from Ileret, Kenya. *Science* 323: 1197–1201.

Coqueugniot, H., J.-J. Hublin, F. Veillon, F. Houët, T. Jacob. 2004. Early brain growth in *Homo erectus* and implications for cognitive ability. *Nature* 431: 299–302.

Dean, C., M. G. Leakey, D. Reid, F. Schrenk, G. T. Schwartz, C. Stringer, A. Walker. 2001. Growth processes in teeth distinguish modern humans from *Homo erectus* and earlier hominins. *Nature* 414: 628–631.

Dean, M. C., B. H. Smith. 2009. Growth and development of the Nariokotome Youth, KNM-WT 15000. In Grine, F. E. et al. (eds.). *The First Humans: Origin and Early Evolution of the Genus* Homo. Heidelberg: Springer, 101–120.

Dobzhansky, T. 1944. On species and races of living and fossil man. *Amer. Jour. Phys. Anthropol.* 2: 251–265.

Dubois, E. 1894. Pithecanthropus erectus, *eine menschenähnliche Uebergangsform aus Java*. Batavia: Landesdruckerei.

Eldredge, N. 1985. *Unfinished Synthesis: Biological Hierarchies and Modern Evolutionary Thought*. New York: Oxford University Press.

Goldschmidt, R. B. 1940. *The Material Basis of Evolution*. New Haven, CT: Yale University Press.

Graves, R. R., A. C. Lupo, R. C. McCarthy, D. J. Wescott, D. L. Cunningham. 2010. Just how strapping was KNM-WT15000? *Jour. Hum. Evol.* 59: 542–554.

Khaitovich, O., I. Hellmann, W. Enard, K. Nowick, M. Leinweber, H. Franz, G. Weiss, M. Lachmann, S. Pääbo. 2005. Parallel patterns of evolution in the genomes and transcriptomes of humans and chimpanzees. *Science* 309: 1850–1854.

Leakey, L. S. B., P. V. Tobias, J. R. Napier. 1964. A new species of *Homo* from Olduvai Gorge. *Nature* 202: 7–9.

Leakey, M. G., F. Spoor, F. H. Brown, P. N. Gathogo, L. N. Leakey, I. McDougall. 2001. New hominin genus from eastern Africa shows diverse middle Pliocene lineages. *Nature* 410: 433–440.

Leakey, R. E. F. 1973. Evidence for an advanced Plio-Pleistocene hominid from East Rudolf, Kenya. *Nature* 242: 447–450.

Leakey, R. E. F. 1976. Hominids in Africa. *Amer. Scientist* 64: 164–178.

Maclarnon, A. M., G. P. Hewitt. 1999. The evolution of human speech: The role of enhanced breathing control. *Amer. Jour. Phys. Anthropol.* 109: 341–363.

Mayr, E. 1950. Taxonomic categories in fossil hominids. *Cold Spring Harbor Symp. Quant. Biol.* 15: 109–118.

Oakley, K. P. 1949. *Man the Tool-Maker*. London: British Museum.

Peichel, C. K., K. S. Nereng, K. A. Ohgl, B. L. E. Cole, P. F. Colosimo, C. A. Buerkle, D. Schluter, D. M. Kingsley. 2001. The genetic architecture of divergence between threespine stickleback species. *Nature* 414: 901–905.

Pilbeam, D. R., E. L. Simons. 1965. Some problems of hominid classification. *Amer. Scientist* 53: 237–259.

Robinson, J. T. 1965. *Homo 'habilis'* and the australopithecines. *Nature* 205: 121–124.

Schwartz, J. H., I. Tattersall. 2005. *The Human Fossil Record, Vol. 3: Genera Australopithecus, Paranthropus, Orrorin, and Overview.* New York: Wiley-Liss, 1634–1653.

Tattersall, I. 2007. *Homo ergaster* and its contemporaries. In W. Henke, I. Tattersall (eds.). *Handbook of Paleoanthropology, Vol. 3.* Heidelberg: Springer.

Tattersall, I. 2009. *The Fossil Trail: How We Know What We Think We Know about Human Evolution.* 2nd ed. New York: Oxford University Press.

Walker, A. C., R. E. F. Leakey. 1993. *The Nariokotome* Homo erectus *skeleton.* Cambridge, MA: Harvard University Press.

Wood, B., M. Collard. 1999. The human genus. *Science* 284: 65–71.

CHAPTER 6: LIFE ON THE SAVANNA

See Aiello and Wheeler (1995) for the "expensive tissue hypothesis" (guts and brain). Body and pubic lice were investigated by Reed et al. (2007). The putative early importance of aquatic resources is discussed by contributions in Cunnane and Stewart (2010). Evidence of fire was reported from Swartkrans by Brain and Sillen (1988) and from Chesowanja by Gowlett et al. (1981). The argument for very early hominid use of fire has been laid out in most detail by Wrangham (2009). Sandgathe et al. (2011) have made the opposite case, that habitual use of fire came very late. Lack of prosociality in chimpanzees has been demonstrated by, among others, Silk et al. (2005). A particularly interesting review of the Oldowan is by Plummer (2004), and raw material transport at Kanjera is analyzed by Braun et al. (2008).

Aiello, L., P. Wheeler. 1995. The expensive-tissue hypothesis: The brain and the digestive system in human and primate evolution. *Curr. Anthropol.* 36: 199–221.

Brain, C. K., A. Sillen. 1988. Evidence from the Swartkrans cave for the earliest use of fire. *Nature* 336: 464–466.

Braun, D. R., T. Plummer, P. Ditchfield, J. V. Ferrari, D. Maina, L. C. Bishop, R. Potts. 2008. Oldowan behavior and raw material transport: Perspectives from the Kanjera Formation. *Jour. Archaeol. Sci.* 35: 2329–2345.

Cunnane, S. C., K. M. Stewart (eds.). 2010. *Human Brain Evolution: The Influence of Freshwater and Marine Food Resources.* Hoboken, NJ: Wiley-Blackwell.

Gowlett, J. A. J., J. W. K. Harris, D. Walton, B. A. Wood. 1981. Early archaeological sites, hominid remains and traces of fire from Chesowanja, Kenya. *Nature* 294: 125–129.

Plummer, T. 2004. Flaked stones and old bones: Biological and cultural evolution at the dawn of technology. *Yrbk Phys. Anthropol.* 47: 118–164.

Reed, D. L., J. E. Light, J. M. Allen, J. J. Kirchman. 2007. Pair of lice lost or paradise regained: The evolutionary history of anthropoid primate lice. *BMC Biol.* 5:7 doi: 10.1186/1741–7007–5–7.

Sandgathe, D. M., H. L. Dibble, P. Goldberg, S. P. McPherron, A. Turq, L. Niven, J. Hodgkins. 2011. Timing of the appearance of habitual fire use. *Proc. Natl Acad. Sci. USA,* doi/10.173/pnas.1106759108.

Silk, J. B., S. F. Brosnan, J. Vonk, D. J. Povinelli, A. S. Richardson, S. P. Lambeth, J. Mascaro, S. J. Schapiro. 2005. Chimpanzees are indifferent to the welfare of unrelated group members. *Nature* 437: 1357–1359.

Wrangham, R. 2009. *Catching Fire: How Cooking Made Us Human.* New York: Basic Books.

CHAPTER 7: OUT OF AFRICA . . . AND BACK

The first Dmanisi hominid was described by Gabunia and Vekua (1995), and later ones by Gabunia et al. (2000a,b), Gabounia et al. (2002), de Lumley and Lordkipanidze (2006) and Lordkipanidze (2007); for dating see de Lumley et al. (2002). The toothless Dmanisi skull was interpreted by Lordkipanidze et al. (2005), and for environmental reconstruction see Messager et al. (2010). An up-to-date review of handaxe cultures (and indeed all ancient stone tool making traditions) can be found in Klein (2009). The Olorgesailie hominid and tool assemblage was described by Potts et al. (2004), and the Isimila site by Howell et al. (1972). Earliest Acheulean was reported by Lepre et al. (2011). Holloway et al. (2004) list fossil hominid brain sizes, and describe endocasts. The Buia hominid was described by Abbate et al. (1998); the Daka specimen by Asfaw et al. (2002); and the two lineages at Ileret by Spoor et al. (2007). Brown et al. (2004) described *Homo floresiensis;* for additional discussion see Martin et al. (2006) and Jungers and Baab (2009), and bibliographies therein.

Abbate, E., A. Albianelli, A. Azzaroli, M. Benvenuti, B. Tesfamariam, P. Bruin, N. Cipriani, R. J. Clarke, G. Ficcarelli, R. Macchiarelli, G. Napoleone, M. Papini, L. Rook, M. Sagri, T. M. Tecle, D. Torre, I. Villa. 1998. A one-million-year-old *Homo* cranium from the Danakil (Afar) Depression of Eritrea. *Nature* 393: 458–460.

Asfaw, B., W. H. Gilbert, Y. Beyene, W. K. Hart, P. R. Renne, G. WoldeGabriel, E. S. Vrba, T. D. White. 2002. Remains of *Homo erectus* from Bouri, Middle Awash, Ethiopia. *Nature* 416: 317–320.

Brown, P., T. Sutikna, M. J. Morwood, R. P. Soejono, Jatmiko, E. W. Saptomo, R. A. Due. 2004. A new small-bodied hominin from the Late Pleistocene of Flores, Indonesia. *Nature* 431: 1055–1061.

de Lumley, H., D. Lordkipanidze, G. Féraud, T. Garcia, C. Perrenoud, C. Falguères, J. Gagnepain, T. Saos, P. Voinchet. 2002. Datation par la méthode ^{40}Ar/^{39}Ar de la couche de cendres volcaniques (couche VI) de Dmanissi (Géorgie) qui a livré des restes d'hominidés fossils de 1.81 Ma. *C. R. Palévol.* 1: 181–189.

Gabounia, Léo, M-A. de Lumley, A. Vekua, D. Lordkipanidze, H. de Lumley. 2002. Découverte d'un nouvel hominidé à Dmanissi (Transcaucasie, Géorgie). *C. R. Palevol* 1: 243–253.

Gabunia L., Vekua A. 1995. A Plio-Pleistocene hominid from Dmanisi, east Georgia, Caucasus. *Nature* 373: 509–512

Gabunia L., Vekua A., Lordkipanidze D. 2000a. The environmental contexts of early human occupations of Georgia (Transcaucasia). *Jour. Hum. Evol.* 38: 785–802

Gabunia L., Vekua A., Lordkipanidze D., Swisher C. C., Ferring R., Justus A., Nioradze M., Tvalcrelidze M., Anton S., Bosinski G. C., Jöris O., de Lumley M. A., Majusuradze G., Mouskhelishvili A. 2000b. Earliest Pleistocene hominid cranial remains from Dmanisi, Republic of Georgia: Taxonomy, geological setting and age. *Science* 288: 1019–1025

Holloway, R. L., D. C. Broadfield, M. S. Yuan. 2004. *The Human Fossil Record, Vol. 3: Hominid Endocasts: The Paleoneurological Evidence.* New York: Wiley-Liss.

Jungers, W. L., K. Baab. 2009. The geometry of hobbits: *Homo floresiensis* and human evolution. *Significance* 6: 159–164.

Klein, R. 2009. *The Human Career: Human Biological and Cultural Origins,* 3rd ed. Chicago: University of Chicago Press.

Lepre, C. J., H. Roche, D. V. Kent, S. Harmand, R. L. Quinn, J.-P. Brugal, P.-J. Texier, A. Lenoble, C. S. Feibel. 2011. An earlier age for the Acheulian. *Nature* 477: 82–85.

Lordkipanidze, D., A. Vekua, R. Ferring, G. P. Rightmire, J. Agusti, G. Kiladze, A. Mouskhelishvili, M. Ponce de Leon, M. Tappen, C. P. E. Zollikofer. 2005. The earliest toothless hominin skull. *Nature* 434: 717–718.

Lordkipanidze, D., T. Jashashvili, A. Vekua, M. Ponce de Leon, C. P. E. Zollikofer, G. P. Rightmire, H. Pontzer, R. Ferring, O. Oms, M. Tappen, M. Bukhsianidze, J. Agusti, R. Kahlke, G. Kiladze, B. Martinez-Navarro, A. Mouskhelishvili, M. Nioradze, L. Rook. 2007. Postcranial evidence from early *Homo* from Dmanisi, Georgia. *Nature* 449: 305–310.

Martin, R. D., M. MacLarnon, J. L. Phillips, W. B. Dobyns. 2006. Flores hominid: New species or microcephalic dwarf? *Anat. Rec.* 288A: 1123–1145.

Messager, E., V. Lebreton, L. Marquez, E. Russo-Ermoli, R. Orain, J. Renault-Miskovsky, D. Lordkipanidze, J. Despriée, C. Peretto, M. Arzarello. Palaeoenvironments of early hominins in temperate and Mediterranean Eurasia: New palaeobotanical data from Palaeolithic key-sites and synchronous natural sequences. *Quat. Sci. Revs* 30: 1439–1447.

Potts, R., A. K. Behrensmeyer, A. Deino, P. Ditchfield, J. Clark. 2004. Small mid-Pleistocene hominin associated with Acheulean technology. *Science* 305: 75–78.

Spoor, F., M. G. Leakey, P. N. Gathogo, F. H. Brown, S. C. Anton, I. McDougall, C. Kiarie, F. K. Manthi, L. N. Leakey. 2007. Implications of new early *Homo* fossils from Ileret, east of Lake Turkana, Kenya. *Nature* 448: 688–691.

CHAPTER 8: THE FIRST COSMOPOLITAN HOMINID

The Mauer jaw was dated by Wagner et al. (2010). See Tattersall (2009) for background to the various *Homo heidelbergensis* fossils. The Terra Amata site was described by de Lumley and Boone (1976), and the Schoeningen finds by Thieme (1997). See Johnson and McBrearty (2010) for early blade production in Kenya, Marshack (1996) for a description of the Berekhat Ram "Venus," and Thompson (2004) for the potentially early ostrich eggshell beads from Loiyalangani.

de Lumley H., Y. Boone. 1976. Les structures d'habitat au Paléolithique inférieur. In H de Lumley (ed.). *La Préhistoire française vol. 1.* Paris, CNRS, 635–643.

de Lumley, M-A., D. Lordkipanidze. 2006. L'homme de Dmanissi (*Homo georgicus*), il y a 1 810 000 ans. *Paléontologie humaine et Préhistoire* 5: 273–281.

Howell, F. C., G. H. Cole, M. R. Kleindienst, B. J. Szabo, K. P. Oakley. 1972. Uranium-series dating of bone from Isimila prehistoric site, Tanzania. *Nature* 237: 51–52.

Johnson, C. R., S. McBrearty. 2010. 500,000 year old blades from the Kapthurin Formation, Kenya. *Jour. Hum. Evol.* 58: 193–200.

Marshack, A. 1996. A Middle Paleolithic symbolic composition from the Golan Heights: The earliest depictive image. *Curr. Anthropol.* 37: 357–365.

Tattersall, I. 2009. *The Fossil Trail: How We Know What We Think We Know about Human Evolution.* 2nd ed. New York: Oxford University Press.

Thieme H. 1997. Lower Palaeolithic hunting spears from Germany. *Nature* 385: 807–810.

Wagner, G. A., M. Krbetschek, D. Degering, J.-J. Bahain, Q. Shao, C. Falguères, P. Voinchet, J.-M. Dolo, T. Garcia, G. P. Rightmire. 2010. Radiometric dating of the type-site for *Homo heidelbergensis* at Mauer, Germany. *Proc. Nat. Acad. Sci. USA*, doi/10.1073/pnas.1012722107.

CHAPTER 9: ICE AGES AND EARLY EUROPEANS

Van Andel (1994) offers engaging insights into Ice Age–related geology; Vrba (1993, 1996) discusses Plio-Pleistocene environments and faunal turnover pulses from a South African perspective, and Behrensmeyer et al. (1997) present an alternative view based on an East African record. Numerous articles in Delson et al. (2000) deal with Pleistocene geology and faunal change. Important ice core data were presented by EPICA (2004) and discussed by McManus (2004); for a review of sea-floor core data see contributions in Gradstein et al. (2005). The Gran Dolina hominid was described by Carbonell et al. (2008), and *Homo antecessor* by Bermudez de Castro et al. (1997). The evidence for cannibalism at the Gran Dolina was reported by Fernandez-Jalvo et al. (1999), and reviewed by Carbonell et al. (2010). The Sima de los Huesos fossils are most comprehensively described in contributions to Arsuaga et al. (1997), and the latest dating is by Bischoff et al. (2007). See Andrews and Fernandez Jalvo (1997) for a dissenting view on the Sima accumulation. The potential symbolic significance of the Sima handaxe is discussed by Carbonell and Mosquera (2006), and paleoenvironments by Garcia and Arsuaga (2010). For the broader picture of relationships among Middle Pleistocene hominids, see Tattersall and Schwartz (2009).

Andrews, P., Y. Fernadez Jalvo. 1997. Surface modifications of the Sima de los Huesos hominids. *Jour Hum. Evol.* 33: 191–217.

Arsuaga, J.-L., J. M. Bermudez de Castro, E. Carbonell (eds). 1997. Special Issue: The Sima de los Huesos hominid site. *Jour. Hum. Evol.* 33: 105–421.

Behrensmeyer, A. K., N. E. Todd, R. Potts, G. E. McBrinn. 1997. Late Pliocene faunal turnover in the Turkana Basin, Kenya and Ethiopia. *Science* 278: 1589–1594.

Bermudez de Castro, J. M. B, J. L. Arsuaga, E. Carbonell, A Rosas, I. Martínez, M. Mosquera. 1997. A hominid from the Lower Pleistocene of Atapuerca, Spain: Possible ancestor to Neandertals and modern humans. *Science* 276: 1392–1395.

Bischoff, J. L., R. W. Williams, R. J. Rosenbauer, A. Aramburu, J. L. Arsuaga, N. García, G. Cuenca-Bescós. 2007. High-resolution U-series dates from the Sima de los Huesos hominids yields 600±66 kyrs: implications for the evolution of the early Neanderthal lineage. *Jour. Archaeol. Sci.* 34: 763–770.

Carbonell, E., M. Mosquera. 2006. The emergence of symbolic behaviour: The sepulchral pit of Sima de los Huesos, Sierra de Atapuerca, Burgos, Spain. *C. R. Palevol.* 5: 155–160.

Carbonell, E., I. Caceres, M. Lizano, P. Saladie, J. Rosell, C. Lorenzo, J. Vallverdu, R. Huguet, A. Canals, J. M. Bermudez de Castro. 2010. Cultural cannibalism as a paleoeconomic system in the European lower Pleistocene. *Curr. Anth.* 51: 539–549.

Carbonell, E., J. M. Bermudez de Castro, J. M. Pares, A. Perez-Gonzalez, G. Cuenca-Bescos, A. Olle, M. Mosquera, R. Huguet, J. van der Made, A. Rosas, R. Sala, J. Vallverdu, N. Garcia, D. E. Granger, M. Martinon-Torres, X. P. Rodriguez, G. M. Stock, J. M. Verges, E. Allue, F. Burjachs, I. Càceres, A. Canals, A. Benito, C. Diez, M. Lozanao, A. Mateos, M. Navazo, J. Rodriguez, J. Rosell, J. L. Arsuaga. 2008. The first hominin of Europe. *Nature* 452: 465–469.

Delson, E., I. Tattersall, J. A. Van Couvering, A. S. Brooks. 2000. *Encyclopedia of Human Evolution and Prehistory*, 2nd ed. New York: Garland Press.

EPICA community. 2004. Eight glacial cycles from an Antarctic ice core. *Nature* 429: 623–628.

Fernandez-Jalvo, Y., J. Carlos Diez, I. Càceres, J. Rosell. 1999. Human cannibalism in the Early Pleistocene of Europe (Gran Dolina, Sierra de Atapuerca, Burgos, Spain). *Jour. Hum. Evol.* 37: 591–622.

Garcia, N., J.-L. Arsuaga. 2010. The Sima de los Huesos (Burgos, northern Spain): Palaeoenvironment and habitats of *Homo heidelbergensis* during the Middle Pleistocene. *Quat. Sci. Revs.*, doi:10:1016/jquascirev.2010.11 .08.

Gradstein, F., J. Ogg, A. G. Smith (eds). 2005. *A Geological Time Scale 2004*. Cambridge: Cambridge University Press.

McManus, J. F. 2004. A great grand-daddy of ice cores. *Nature* 429: 611–612.

Tattersall, I., J. H. Schwartz. 2009. Evolution of the genus *Homo. Ann. Rev. Earth Planet. Sci.* 37: 67–92.

Van Andel, T. H. 1994. *New Views on an Old Planet*. Cambridge: University of Cambridge Press.

Vrba, E. S. 1993. The pulse that produced us. *Natural History* 102 (5): 47–51.

Vrba, E. S. 1996. *Paleoclimate and Evolution, with Emphasis on Human Origins*. New Haven, CT: Yale University Press.

CHAPTER 10: WHO WERE THE NEANDERTHALS?

For an account of the Biache fossil, see Schwartz and Tattersall (2002); for Reilingen, see Dean et al. (1998). For the coexistence of lineages in Europe see Tattersall and Schwartz (2006), for Finnish Mousterian see Schulz (2000), for the Altai Neanderthal genetic signature see Krause et al. (2007), and for Neanderthal avoidance of periglacial environments see Patou-Mathis (2006). The late presumed Neanderthal occurrence in northern Russia was reported by Slimak et al. (2011). Pearson et al. (2006) discuss Neanderthal climatic adaptation, as do various contributions in Van Andel and Davies (2003).

The first report of Neanderthal mtDNA was by Krings et al. (1997), and a recent report and review is by Briggs et al. (2009). The draft Neanderthal genome was reported by Green et al. (2010), and the Denisova genome by Reich et al. (2010). See Cohen (2010) for a short account of modern interspecific hybrids, Johnson et al. (2006) for lion and tiger ancestry, and Jolly (2001) for the hamadryas and gelada hybrid zone and implications. For a variety of views on the Abrigo do Lagar Velho skeleton see Zilhao and Trinkaus (2002), and for an account of the Peştera cu Oase skull see Trinkaus et al. (2003).

Consult Smith et al. (2010) for the most recent report and synthesis of Neanderthal dental development, Ponce de Leon and Zollikofer (2001) for Neanderthal cranial development, and Gunz et al. (2010) for Neanderthal vs. modern brain development trajectories. For Neanderthal hair and skin color, refer to Lalueza-Fox et al. (2007). The introgression of the *microcephalin* gene variant into *Homo sapiens* from an archaic hominid lineage was suggested by Evans et al. (2006). See Stiner and Kuhn (1992) for a comparison of Neanderthal subsistence at Italian sites. See Richards and Trinkaus (2009) for a summary of nitrogen-isotope studies, Bocherens et al. (2005) for the St.-Césaire nitrogen isotope data and interpretations, and Henry et al. (2010) for the Shanidar and Spy plant microfossil analyses. Lalueza-Fox et al. (2010) present mtDNA data from El Sidrón, while Vallverdú et al. report on site formation and population sizes at Abric Romaní. Quote is from Zimmer (2010). The human bone tool from La Quina was described by Verna and D'Errico (2010).

Bocherens, H. D. G. Drucker, D. Billiou, M. Patou-Mathis, B. Vandermeersch. 2005. Isotopic evidence for diet and subsistence pattern of the Saint-Césaire I Neanderthal: review and use of a multi-source mixing model. *Jour. Hum. Evol.* 49: 71–87.

Briggs, A. W., J. M. Good, R. E. Green, J. Krause, T. Maricic, U. Stenzel, C. Lalueza-Fox and numerous others. 2009. Targeted retrieval and analysis of five Neanderthal mtDNA genomes. *Science* 325: 318–321.

Cohen, J. 2010. *Almost Chimpanzee: Searching for What Makes us Human in Rainforests, Labs, Sanctuaries and Zoos.* New York: Times Books.

Dean, D., J.-J. Hublin, R. Holloway, R. Ziegler. 1998. On the phylogenetic position of the pre-Neandertal specimen from Reilingen, Germany. *Jour. Hum. Evol.* 34: 485–508.

Evans, P. D., M. Mekel-Bobrov, E. J. Vallender, R. R. Hudson, B. T. Lahn. 2006. Evidence that the adaptive allele of the brain size gene *microcephalin* introgressed into *Homo sapiens* from an archaic *Homo* lineage. *Proc. Nat. Acad. Sci. USA* 103: 18178–18183.

Green, R. E., J. Krause, A. W. Briggs, T. Maricic, U. Stenzel, M. Kirchner, N. Patterson and 49 others. 2010. A draft sequence of the Neanderthal genome. *Science* 328: 710–722.

Gunz, P., S. Neubauer, B. Maureille, J.-J. Hublin. 2010. Brain development after birth differs between Neanderthals and modern humans. *Curr. Biol.* 20 (21): R921–R922.

Henry, A. G., A. S. Brooks, D. R. Piperno. 2010. Microfossils in calculus demonstrate consumption of plants and cooked foods in Neanderthal diets (Shanidar III, Iraq; Spy I and II, Belgium). *Proc. Nat. Acad. Sci. USA,* doi/10.1073/pnas.101686108.

Johnson, W. E., E. Eizirik, J. Pecon-Slattery, W. J. Murphy, A. Antunes, E. Teeling, S. J. O'Brien. 2006. The late Miocene radiation of modern Felidae: A genetic assessment. *Science* 311: 73–77.

Jolly, C. J. 2001. A proper study for mankind: Analogies from the papionin monkeys and their implications for human evolution. *Yrbk Phys. Anthropol.* 44: 177–204.

Krause J., Orlando L., Serre D., Viola B., Prüfer K., Richards M. P., Hublin J. J., Hänni C., Derevianko A. P., Pääbo S. 2007. Neanderthals in central Asia and Siberia. *Nature* 449: 1–3.

Lalueza-Fox, C., A. Rosas, A. Estalrrich, E. Gigli, P. F. Campos, A. Garcia-Tabernero, S. Garcia-Vargas and 9 others. 2010. Genetic evidence for patrilocal mating behavior among Neandertal groups. *Proc. Nat. Acad. Sci. USA,* doi/10.1073/pnas.1011533108.

Lalueza-Fox, C., H. Rompler, D. Caramelli, C. Staubert, G. Catalano, D. Hughes, N. Rohland and 10 others. 2007. A melanocortin 1 receptor allele suggests varying pigmentation among Neanderthals. *Science* 318: 1453–1455.

Patou-Mathis, M. 2006. Comportements de subsistance des Néandertaliens d'Europe. In B. Demarsin and M. Otte (eds.). *Neanderthals in Europe*. Liege, ERAUL, 117: 9–14.

Pearson, O. M., R. M. Cordero, A. M. Busby. 2006. How different were the Neanderthals' habitual activities? A comparative analysis with diverse groups of recent humans. In K. Harvati and T. Harrison (eds.). *Neanderthals Revisited: New Approaches and Perspectives*. Berlin: Springer, 135–156.

Ponce de León, M. S. and C. P. E. Zollikofer. 2001. Neanderthal cranial ontogeny and its implications for late hominid diversity. *Nature* 412: 534–538.

Reich, D., R. E. Green, M. Kirchner, J. Krause, N. Patterson, E. Y. Durand, B. Viola and numerous others. 2010. Genetic history of an archaic hominin group from Denisova Cave in Siberia. *Nature* 468: 1053–1060.

Schulz, H.-P. 2000/2001. The lithic industry from layers IV-V, Susiluola Cave, Western Finland. *Prehist. Europ.* 16/17: 43–56.

Schwartz, J. H., I. Tattersall. 2002. *The Human Fossil Record, Vol. 1: Terminology and Craniodental Morphology of Genus Homo (Europe)*. New York: Wiley-Liss.

Slimak, L, J. I. Svendsen, J. Mangerud, H. Plisson, H. P. Heggen, A Brugère, P. Y. Pavlov. 2011. Late Mousterian persistence near the Arctic Circle. *Science* 332: 841–845.

Stiner, M., S. Kuhn. 1992. Subsistence, technology, and adaptive variation in Middle Paleolithic Italy. *Amer. Anthropol.* 94: 306–339.

Tattersall, I., Schwartz, J. H. 2006. The distinctiveness and systematic context of *Homo neanderthalensis*. In K. Harvati and T. Harrison (eds.). *Neanderthals Revisited: New Approaches and Perspectives*. Berlin: Springer, 9–22.

Trinkaus, E., S. Milota, R. Rodrigo, G. Mircea, O. Moldovan. 2003. Early modern human remains from the Peştera cu Oase, Romania. *Jour. Hum. Evol.* 45: 245–253.

Vallverdú, J., M. Vaquero, I. Cáceres, E. Allué, J. Rosell, P. Saladié, G. Chacón, A. Ollé, A. Canals, R. Sala, M. A. Courty, E. Carbonell. 2010. Sleeping Activity Area within the Site Structure of Archaic Human Groups: Evidence from Abric Romaní Level N Combustion Activity Areas. *Curr. Anthropol.* 51: 137–145.

Van Andel, T. H., W. Davies. 2003. *Neanderthals and Modern Humans in the European Landscape during the Last Glaciation* (McDonald Institute Monographs). Oxford, UK: Oxbow Books.

Verna, C., F. D'Errico. 2010. The earliest evidence for the use of human bone as a tool. *Jour. Hum. Evol.* 60: 145–147.

Zilhao, J., E. Trinkaus (eds.). 2002. Portrait of the artist as a child: The Gravettian human skeleton from the Abrigo do Lagar Velho and its Archeological Context. *Trab. Arqueol.* 22: 1–604.

Zimmer, C. 2010. Bones give peek into the lives of Neanderthals. *New York Times*, 20 December.

CHAPTER 11: ARCHAIC AND MODERN

For an excellent general account of the Mousterian, see Klein (2009). Soressi and D'Errico (2007) present an overview of putatively symbolic objects and materials in the Mousterian. Finlayson (2009) provides a fascinating perspective on Neanderthal populations and environments. For the presence of Neanderthals at Arcy, see Hublin et al. (1996), and for recent views on the Châtelperronian at Arcy and St.-Césaire, see Bar-Yosef and Bordes (2010) and Higham et al. (2010). Evidence for early and brief replacement of Neanderthals by moderns was presented by Pinhasi et al. (2011).

Bar-Yosef, O., J.-G. Bordes. 2010. Who were the makers of the Châtelperronian culture? *Jour. Hum. Evol.* 59: 586–593.

Finlayson, C. 2009. *The Humans Who Went Extinct: Why Neanderthals Died Out and We Survived.* Oxford, UK: Oxford University Press.

Higham, T., R. Jacobi, M. Julien, F. David, L. Basell, R. Wood, W. Davies, C. B. Ramsey. 2010. Chronology of the Grotte du Renne (France) and implications for the context of ornaments and human remains within the Châtelperronian. *Proc. Nat. Acad. Sci. USA* 107: 20234–20239.

Hublin, J.-J., F. Spoor, M. Braun, F. Zonneveld, and S. Condemi. 1996. A late Neanderthal associated with Upper Palaeolithic artefacts. *Nature* 381: 224–226.

Klein, R. 2009. *The Human Career,* 3rd ed. Chicago: University of Chicago Press.

Pinhasi, R., T. F. G. Higham, L. V. Golubova, V. B. Doronichev. 2011. Revised age of late Neanderthal occupation and the end of the Middle Paleolithic in the northern Caucasus. *Proc Nat. Acad. Sci. USA* 108: 8611–8616.

Soressi, M., F. D'Errico. 2007. Pigments, gravures, parures: Les comportements symboliques controversés des Néandertaliens. In B. Vandermeersch, B. Maureille (eds.). *Les Néandertaliens: Biologie et Cultures.* Paris: Editions du CTHS, 297–309.

CHAPTER 12: ENIGMATIC ARRIVAL

For the very earliest *Homo sapiens* fossils in Africa, see MacDougall et al. (2005), White et al. (2003) and Clark et al. (2003). For the Middle Stone Age see the review in Klein (2009), and for the Aterian and associated hominids see Balter (2011) and contributions in Garcea (2010) and Hublin and McPherron (2011). Drake et al. (2010) discuss the "green Sahara." For dating of the Levantine sites see Bar-Yosef (1998) and Grün et al. (2005), and Coppa et al. (2005); for the Levantine hominids see Schwartz and Tattersall (2003, 2010).

Tishkoff et al. (2009) provide the most comprehensive recent discussion of African genetic diversity. Campbell and Tishkoff (2010) provide an overview and excellent bibliography. See also commentary in Gibbons (2009), and Scheinfeldt et al. (2010) for synthesis with linguistics and archaeology. For bottlenecking, see Jorde et al. (1998) and Harpending and Rogers (2000); specifically in connection with Mount Toba, see Ambrose (1998); further commentary on the Mount Toba scenario is by Ambrose (2003) and Gathorne-Hardy and Harcourt-Smith (2003). See DeSalle and Tattersall (2008) for molecular techniques and a detailed summary of the molecular evidence for human spread. Liu et al. (2010) describe the alleged ancient *Homo sapiens* jaw from China, and Pitulko et al. (2004) the earliest occupation north of the Arctic Circle.

Ambrose, S. H. 1998. Late Pleistocene human population bottlenecks, volcanic winter, and differentiation of modern humans. *Jour. Hum. Evol.* 34: 623–651.

Ambrose, S. H. 2003. Did the super-eruption of Toba cause a human population bottleneck? Reply to Gathorne-Hardy and Harcourt-Smith. *Jour. Hum. Evol.* 45: 231–237.

Balter, M. 2011. Was North Africa the launch pad for modern human migrations? *Science* 331: 20–23.

Bar-Yosef, Y. 1998. The chronology of the Middle Paleolithic of the Levant. In T. Akazawa, K. Aoki, O. Bar-Yosef (eds.). *Neandertals and Modern Humans in Western Asia*. New York: Plenum Press, 39–56.

Campbell, M., S. A. Tishkoff. 2010. The evolution of human genetic and phenotypic variation in Africa. *Curr. Biol.* 20: R166–R173.

Clark, J. D., Y. Beyene, G. WoldeGabriel, W. K. Hart, P. R. Renne, H. Gilbert, A. Defleur, G. Suwa, S. Katoh, K. R. Ludwig, J.-R. Boisserie, B. Asfaw, T. D. White. 2003. Stratigraphic, chronological and behavioural contexts of Pleistocene *Homo sapiens* from Middle Awash, Ethiopia. *Nature* 423: 747–752.

Coppa, A., R. Grün, C. Stringer, S. Eggins, R. Vargiu. 2005. Newly recognized Pleistocene human teeth from Tabūn Cave, Israel. *Jour. Hum. Evol.* 49: 301–315.

DeSalle, R., I. Tattersall. 2008. *Human Origins: What Bones and Genomes Tell Us about Ourselves*. College Station, TX: Texas A&M University Press.

Drake, N. A., R. M. Blench, S. J. Armitage, C. S. Bristow, K. H. White. 2010. Ancient watercourses and biogeography of the Sahara explain the peopling of the desert. *Proc. Nat. Acad. Sci. USA* 108: 458–462.

Garcea, E. A. A. (ed.). 2010. *South-Eastern Mediterranean Peoples between 130,000 and 10,000 Years Ago*. Oxford, UK: Oxbow Books.

Gathorne-Hardy, F. J., W. E. H. Harcourt-Smith. 2003. The super-eruption of Toba, did it cause a human bottleneck? *Jour. Hum. Evol.* 45: 227–230.

Gibbons, A. 2009. Africans' deep genetic roots reveal their evolutionary story. *Science* 324: 575.

Grün, R., C. Stringer, F. McDermott, R. Nathan, N. Porat, S. Robertson, L. Taylor, G. Mortimer, S. Eggins, M. McCulloch. 2005. U-series and ESR analyses of bones and teeth relating to the human burials from Skhūl. *Jour. Hum. Evol.* 49: 316–334.

Harpending, H., A. R. Rogers. 2000. Genetic perspectives on human origins and differentiation. *Ann. Rev. Genom. Hum. Genet.* 1: 361–385.

Hublin, J. J., S. McPherron. 2011. *Modern Origins: A North African Perspective*. New York: Springer.

Klein, R. 2009. *The Human Career*, 3rd ed. Chicago: University of Chicago Press.

Liu, W., C.-Z. Jin, Y.-Q. Zhang, Y.-J. Cai, S. Zing, X.-J. Wu, H. Cheng and 6 others. 2010. Human remains from Zhirendong, South China, and modern human emergence in East Asia. *Proc. Nat. Acad. Sci. USA* 107: 19201–19206.

McDougall, I., F. H. Brown, J. G. Fleagle. 2005. Stratigraphic placement and age of modern humans from Kibish, Ethiopia. *Nature* 433: 733–736.

Pitulko, V. V., P. A. Nikolsky, E. Y. Girya, A. E. Basilyan, V. E. Tumskoy, S. A. Koulakov, S. N. Astakhov, E. Y. Pavlova, M. A. Anisimov. 2004. The Yana RHS site: Humans in the Arctic before the Last Glacial Maximum. *Science* 303: 52–56.

Scheinfeldt, L. B., S. Soi, S. A. Tishkoff. 2010. Working toward a synthesis of archaeological, linguistic and genetic data for inferring African population history. *Proc. Nat. Acad. Sci. USA* 107 (Supp. 2): 8931–8938.

Schwartz, J. H., I. Tattersall. 2010. Fossil evidence for the origin of *Homo sapiens. Yrbk. Phys. Anthropol.* 53: 94–121.

Tishkoff, S. A., F. A. Reed, F. B. Friedlander, C. Ehret, A. Ranciaro. A. Froment, J. B. Hirbo and numerous others. 2009. The genetic structure and history of Africans and African Americans. *Science* 324: 1035–1044.

White, T. D., B. Asfaw, D. DeGusta, H. Gilbert, G. D. Richards, G. Suwa, F. C. Howell. 2003. Pleistocene *Homo sapiens* from Middle Awash, Ethiopia. *Nature* 423: 742–747.

CHAPTER 13: THE ORIGIN OF SYMBOLIC BEHAVIOR

The Skhūl and Oued Djebbana beads were reported by Vanhaeren et al. (2006), the Skhūl pigments were analyzed by D'Errico et al. (2010), and further bead evidence in the North African Aterian was reported by Bouzouggar et al. (2007) and d'Errico et al. (2009). The Blombos plaques were described by Henshilwood et al. (2002), and the beads from the site by Henshilwood et al. (2004). Marean et al. (2007) reported pigments and shellfishing at Pinnacle Point; heat treatment of silcrete there was described by Brown et al. (2009), and pressure-flaking at Blombos by Mourre et al. (2010). For background on the Klasies River Mouth sites, see Deacon and Deacon (1999). The Diepkloof ostrich eggshell containers are described by Texier et al. (2020), and the Enkapune Ya Muto beads by Ambrose (1998); see Mellars (2006) for a discussion of early human dispersal through Eurasia, and Kuhn et al. (2001) for shell beads from Lebanese and Turkish sites.

Ambrose, S. H. 1998. Chronology of the later Stone Age and food production in East Africa. *Jour. Archaeol. Sci.* 25: 377–392.

Bouzouggar, A., N. Barton, M. Vanhaeren, F. d'Errico, S. Colcutt, T. Higham, E. Hodge and 8 others. 2007. 82,000-year-old shell beads from North Africa and implications for the origins of modern human behavior. *Proc. Nat. Acad. Sci. USA* 104: 9964–9969.

Brown, K. S., C. W. Marean, A. I. R. Herries, Z. Jacobs, C. Tribolo, D. Braun, D. L. Roberts, M. C. Meyer, J. Bernatchez. 2009. Fire as an engineering tool of early modern humans. *Science* 325: 859–862.

Deacon, H., J. Deacon. 1999. *Human beginnings in South Africa: Uncovering the Secrets of the Stone Age.* Cape Town: David Philip.

d'Errico, F., M. Vanhaeren, N. Barton, A. Bouzouggar, H. Mienis, D. Richter, J.-J. Hublin, S. P. McPherron, P. Lozouet. 2009. Additional evidence on the use of personal ornaments in the Middle Paleolithic of North Africa. *Proc. Nat. Acad. Sci. USA* 106: 16051–16056.

d'Errico, F., H. Salomon, C. Vignaud, C. Stringer. 2010. Pigments from Middle Paleolithic leves of es-Skhūl (Mount Carmel, Israel). *Jour. Archaeol. Sci.* 37: 3099–3110.

Henshilwood, C., F. d'Errico, M. Vanhaeren, K. van Niekerk, Z. Jacobs. 2004. Middle Stone Age shell beads from South Africa. *Science* 304: 404.

Henshilwood, C. S., F. d'Errico, R. Yates, Z. Jacobs, C. Tribolo, G. A. T. Duller, N. Mercier and 4 others. 2002. Emergence of modern human behavior: Middle Stone Age engravings from South Africa. *Science* 295: 1278–1280.

Kuhn, S., M. C. Stiner, D. S. Reese, E. Gulec. 2001. Ornaments of the earliest Upper Paleolithic: New Insights from the Levant. *Proc. Nat. Acad. Sci. USA* 98: 7641–7646.

Marean, C. W., M. Bar-Matthews, J. Bernatchez, E. Fisher, P. Goldberg, A. I. R. Herries, Z. Jacobs and 7 others. 2007. Early use of marine resources and pigment in South Africa during the Middle Pleistocene. *Nature* 449: 905–908.

Mellars, P. 2006. Going east: New genetic and archaeological perspectives on the modern human colonization of Eurasia. *Science* 313: 796–800.

Mourre, V., P. Villa, C. S. Henshilwood. 2010. Early use of pressure flaking on lithic artifacts at Blombos Cave, South Africa. *Science* 330: 659–662.

Texier P. J., G. Porraz, J. Parkington J.-P. Rigaud, C. Poggenpoel, C. Miller, C. Tribolo, C. Cartwright, A. Coudenneau, R. Klein, T. Steele, C. Verna. 2010. A Howiesons Poort tradition of engraving ostrich eggshell containers dated to 60,000 years ago at Diepkloof Rock Shelter, South Africa. *Proc. Nat. Acad. Sci. USA.* 107: 6180–6185.

Vanhaeren, M., F. d'Errico, C. Stringer, S. L. James, J. A. Todd, H. K. Mienis. 2006. Middle Paleolithic shell beads in Israel and Algeria. *Science* 312: 1785–1788.

CHAPTER 14: IN THE BEGINNING WAS THE WORD

For the implication of the FOXP2 gene in speech disorders, see Lai et al. (2001), and for its identification in Neanderthals see Krause et al. (2007). For a discussion of the larynx, facial proportions, and speech see P. Lieberman (2007) and D. E. Lieberman (2011). For amusing advocacy of theory of mind see Dunbar (2004), and for the role of language in symbolic thought see Tattersall (2008). Atkinson (2011) discusses the potential significance of phonemic diversity. For Nicaraguan sign language see Kegl et al. (1999), and for the case of Ildefonso see Schaller (1991). Jill Bolte Taylor (2006) describes the effects of her stroke on her language abilities. DeSalle and Tattersall (2011) provide an account of the functioning and long history of the human brain, and Geschwind (1966) discusses the putative significance of the angular gyrus. Coolidge and Wynn (2009) and Balter (2010) discuss working memory.

Atkinson, Q. D. 2011. Phonemic diversity supports a serial founder effect model of language expansion from Africa. *Science* 332: 346–349.

Balter, M. 2010. Did working memory spark creative culture? *Science* 328: 160–163.

Coolidge, F. L., T. Wynn. 2009. *The Rise of* Homo sapiens: *The Evolution of Modern Thinking.* New York: Wiley-Blackwell.

DeSalle, R., I. Tattersall. 2011. *Brains: Big Bangs, Behavior and Beliefs.* New Haven, CT: Yale University Press.

Dunbar, R. I. M. 2004. *The Human Story: A New History of Mankind's Evolution.* London: Faber & Faber.

Geschwind, N. 1964. The development of the brain and the evolution of language. *Monogr. Ser. Lang. Ling.* 17: 155–169.

Jorde, L. B., M. Bamshad, A. R. Rogers. 1998. Using mitochondrial and nuclear DNA markers to reconstruct human evolution. *BioEssays* 20: 126–136.

Kegl, J., A. Senghas, M. Coppola. 1999. Creation through contact: Sign language emergence and sign language change in Nicaragua. In M. deGraaf (ed.). *Comparative Grammatical Change: The Intersection of Language Acquisition, Creole Genesis and Diachronic Syntax.* Cambridge, MA: MIT Press, 179–237.

Klein, R. 2009. *The Human Career,* 3rd ed. Chicago: University of Chicago Press.

Krause, J., C. Lalueza-Fox, L. Orlando, W. Enard, R. E. Green, H. A, Burbano, J.-J. Hublin and 6 others. 2007. The derived *FOXP2* variant of modern humans was shared with Neandertals. *Curr. Biol.* 17: 1908–1912.

Lai, C. S., S. E. Fisher, J. A, Hurst, F. Vargha-Khadem, A. P. Monaco. 2001. A forkhead-domain gene is mutated in a severe speech and language disorder. *Nature* 413: 519–523.

Lieberman, D. E. 2011. *The Evolution of the Human Head.* Cambridge, MA: Harvard University Press.

Lieberman, P. 2007. The evolution of human speech: Its anatomical and neural bases. *Curr. Anthropol.* 48: 39–66.

Ohnuma, K., K. Aoki, T. Akazawa. 1997. Transmission of tool-making through verbal and non-verbal communication: Preliminary experiments in Levallois flake production. *Anthropol. Sci.* 105 (3): 159–168.

Schaller, S. 1991. *A Man without Words.* New York: Summit Books.

Schwartz, J. H., I. Tattersall. 2003. *The Human Fossil Record, Vol 2: Craniodental Morphology of Genus* Homo *(Africa and Asia).* New York: Wiley-Liss.

Tattersall, I. 2008. An evolutionary framework for the acquisition of symbolic cognition by *Homo sapiens. Comp. Cogn. Behav. Revs* 3: 99–114.

Taylor, J. B. 2006. *My Stroke of Insight: A Brain Scientist's Personal Journey.* New York: Viking.

Marcus (2008) entertainingly details the deficiencies of the human mind. Earth's erosional history is discussed by Wilkinson (2005), and the genetic underpinnings of violence and their neural correlates by Meyer-Lindburg et al. (2006). Crutzen (2002) summarized justifications for the Anthropocene.

Crutzen, P. 2002. Geology of mankind. *Nature* 415: 23.

Marcus, G. 2008. *Kluge: The Haphazard Evolution of the Human Mind.* New York: Houghton Mifflin.

Meyer-Lindburg, A., J. W. Buckholtz, B. Kolachana, A. R. Hariri, L. Pezawas, G. Blasi, A. Wabnitz and 6 others. 2006. Neural mechanisms of genetic risk for impulsivity and violence in humans. *Proc. Nat. Acad. Sci. USA* 103: 6269–6274.

Wilkinson, B. H. 2005. Humans as geologic agents: A deep-time perspective. *Geology* 33 (3): 161–164.

INDEX